W9-AXH-383

Siena College Library
Loudonville, N.Y.

Atoms, molecules and chemical change

PRENTICE-HALL INTERNATIONAL, INC. *London*
PRENTICE-HALL OF AUSTRALIA PTY. LTD. *Sydney*
PRENTICE-HALL OF CANADA, LTD. *Toronto*
PRENTICE-HALL OF INDIA PRIVATE LIMITED *New Delhi*
PRENTICE-HALL OF JAPAN, INC. *Tokyo*

Atoms, molecules and chemical change

Third Edition

RUSSELL H. JOHNSEN
Professor of Chemistry, The Florida State University

ERNEST GRUNWALD
Professor of Chemistry, Brandeis University

PRENTICE-HALL, INC.
Englewood Cliffs, N. J.

SIENA COLLEGE LIBRARY
LOUDONVILLE, N. Y. 12211

Current printing (last digit):
10 9 8 7 6 5 4 3 2 1

© 1971, 1965, 1960 BY PRENTICE-HALL, INC., ENGLEWOOD CLIFFS, N.J.

All rights reserved. No part of this book may be reproduced in any form, by mimeograph or any other means, without permission in writing from the publisher.

Library of Congress Catalog Card Number: 70–149168

PRINTED IN THE UNITED STATES OF AMERICA
13-050260-X

Designed by Betty Binns
Illustration by Felix Cooper

Cover design after an ancient Mexican motif.

QD
453.2
.J64
1971

Preface

TO THE THIRD EDITION

ATOMS, MOLECULES and CHEMICAL CHANGE is intended to be a textbook of chemistry for general education. We have tried to achieve a point of view in which chemistry emerges as a subdivision of natural philosophy rather than of technology. Many courses in general chemistry nowadays take the existence of atoms and molecules for granted at the outset, so that time might be gained in which to develop the technical consequences of these concepts. We feel, however, that such an approach is ill suited for general education. Our book traces the gradual development of modern ideas concerning atomic and molecular structure, beginning with the operational definition of a pure substance and ending with modern theories of the nucleus. Throughout the book we emphasize the interplay of theory and experiment and lead from one to the other. We hope that students of this book will end up with a good idea of the experimental and logical foundations of chemistry.

In preparing the third edition we have tried to keep the original approach substantially intact. However, many sections have been rewritten to improve the accuracy or clarity or to bring the material up-to-date. The chapters on Bohr's theory of the hydrogen atom and on covalent bonding have been revised extensively, and a brief chapter on stoichiometry and chemical equations has been added.

The material in this book is suitable for a one-semester chemistry course or (by careful selection of topics) for a one-quarter course. The book has also been used as a text for a one-quarter course devoted primarily to atomic structure, by selection of chapters 1 through 4 (quickly), 5, 8 through 11, 17 and 18.

We thank our wives for their continued encouragement and their help in the preparation of the manuscript.

R.H.J. / E.G.

Acknowledgments

The authors wish to express thanks for permission to quote from the following copyright works:

Max Born, *Experiment and Theory in Physics,* on page 12, by permission of Dover Publications, Inc., New York, © 1956.

James Gould Cozzens, *Morning, Noon and Night,* on page 122, by permission of Harcourt, Brace and World, © 1968.

Albert Einstein, *Essays in Science,* on page 222, by permission of Philosophical Library, New York, © 1954.

Louis P. Hammett, *Physical Organic Chemistry,* on page 242, by permission of McGraw-Hill Book Company, © 1940.

W. Heisenberg, *The Physicist's Conception of Nature,* translated from the German by A. J. Pomerans, on page 180, by permission of Harcourt, Brace, & World, Inc., © 1958.

Alfred North Whitehead, *Science and the Modern World,* on pages 28 and 140, by permission of The Macmillan Company, © 1925.

Contents

Atoms, molecules and chemical change

The scientist does not study nature because it is useful. He studies it because he delights in it, and he delights in it because it is beautiful.

HENRI POINCARE (1854–1912)

One

INTRODUCTION

CHEMISTRY IS the science that deals with the structure and composition of matter and the transformations it undergoes. It is a field of study that has engaged the attention of many people over many years, not only because of the great complexity and variety of matter, but also because of the great perfection with which the chemist wishes to understand it. Chemistry tries to answer such questions as these: What is the make-up or composition of this particular kind of matter? Why does it have certain qualities and not others? Will it undergo changes into new, and perhaps more interesting kinds of matter? Can it be produced through the transformation of other, more plentiful kinds of matter?

Above all, the chemist is interested in the *structure* of matter. He believes that each kind of matter, if it could be examined with sufficient magnification, would be found to have a characteristic structure; and that if enough were known about the detailed structure for enough different kinds of matter, all the individual qualities of matter could be understood. This belief must, in a large measure, rely on faith, because the kind of structure that is envisaged is best described as *submicroscopic:* it is so fine that even the most powerful of microscopes will not render it visible in any direct sense. Nevertheless, it has been possible to derive a remarkably detailed picture of the structure of matter by a variety of indirect means.

A chemist trying to learn about the structure of matter is not unlike a detective on the scene of a crime. He obtains a clue from this set of observations, and another one from that; then he puts these clues together and builds up a circumstantial case. There is one important difference, however. Whereas the detective's reconstruction of the crime involves real persons who can be questioned, and objects that can be observed, the chemist's picture of the structure of matter can never be verified by direct recourse to the human senses. His case must remain forever circumstantial.

The chemist does the best he can to elucidate the structure of matter within the framework set by these limitations. He tries to make his circumstantial case so airtight, to arrive at a unified picture of the structure of matter from so many different kinds of evidence, that no one but a confirmed skeptic could entertain any serious doubt about the picture. This is an ambitious goal. Clues must be used wherever they are found. While much of the information concerning the structure of matter comes from the work of chemists on the composition and reactions of individual kinds of matter, an equally important share comes from the work of physicists, whose business it is to discover and study such aspects as are common to *all* matter. Thus, physics includes such studies as the motion of objects under the influence of a force, the behavior of matter after electrification or magnetization, or the flow of heat from one portion of matter to another when there is a difference in temperature.

The two sciences, chemistry and physics, complement one another. The former tends to concern itself with the specific characteristics of individual kinds of matter, and the latter with general properties shared by all matter. It is not uncommon, therefore, to find chemists and physicists working on different aspects of the same general problem.

Chemistry's role in today's society

The fruits of these efforts of the chemist are many and varied and have had considerable impact on the way in which we live our lives. Much of this impact has been beneficial. We can cite, for example, the plastics industry, or the new types of metallic alloys that have had so much to do with the successful exploration of space, or the types of material called semiconductors that are basic to the manufacture of transistors and have made possible today's large and fast computers. Medicinal chemistry has provided many cures and prevented much disease. Even the aesthetic side of life has benefited from the efforts of the chemist:

new dyes, new colors and new materials have provided the artist with a variety of media undreamed of twenty years ago. The list is large and has been recited often.

Every coin has two sides, however, and with these benefits have come some problems. Plastic packaging materials are convenient and inexpensive; yet their wide use has increased the amount of persistent litter along our highways. New medicinal drugs have produced unanticipated side effects in some patients, and some drugs have had to be taken off the market. Many products developed for peaceful purposes can also be used as instruments for war. Thus, weed-killers used in agriculture have been used as defoliants in jungle warfare, to the chagrin of many chemists.

A spectacular example of "the other side of the coin" is provided by the pesticide, DDT. After the incredible effectiveness of this substance was discovered in 1939, DDT was hailed as a saviour of mankind—it was used to wipe out the malaria mosquito and to boost food production by controlling agricultural pests in many parts of the world. However, today it is obvious that the wide-spread and uncontrolled use of DDT has placed many animal species in danger of extinction and has given man a burden of toxic material that could, under extreme conditions, endanger his life.

Pollution of air and water has now reached a critical level. Chemical and other industries contribute to this directly. Chemical fertilizers that are washed into streams and lakes, and sewage discharged into streams, contribute another type of pollution—one which promotes the growth of algae at the expense of other life and ultimately fouls the water. Automobile exhausts produce a large fraction of the total air pollutants in the U.S. The increase in air and water pollution is the result of increased industrialization and of a growing population (which is, in turn, a result of the greater longevity brought about by scientific technology).

Fortunately, the pollution problems created by technology can be solved by technology—though usually at some cost. Chemistry will undoubtedly play a role in solving these problems. For example, in the case of pollution by automobiles, one can foresee the possibilities of creating better combustion engines or of powering cars electrically (perhaps using new hydrogen-oxygen fuel cells that produce only water as a waste product). Or consider the problems related to DDT. New insecticides, as well as new methods of pest control, have been developed and will continue to be developed. These pose less of a threat to the balance of nature.

This is the age of the computer, a device that can control machinery and relieve man of many tedious jobs. Chemistry has played a role in its development by perfecting methods for making ultra-pure semiconductor material. What possible undesirable effects will the computer have? How will the computer revolution affect the quality of human life? No one can say for sure, but it is safe to predict that the effects will be profound, and some will be totally unexpected.

Should chemistry, as a science, be blamed for the problems created by technology? The authors argue quite to the contrary: Knowledge of itself is neither good nor bad—it is only the uses to which knowledge is put that can be judged on this scale. The problems that technology has (in some cases, unwittingly) created can be solved, but not by science and technology alone. Essential to the solution of all these problems will be the good sense and social intelligence of mankind.

Submicroscopic models

In the opening paragraphs it was pointed out that the chemist's chief interest is in the *structure* of matter, and that he is concerned with this structure at a submicroscopic level. It must be made completely clear that the submicroscopic picture that the chemist develops is a purely imaginary one: The submicroscopic world and everything in it is a creation of the human mind. This is necessarily so, for the objects that populate the submicroscopic world are much too small to be sensed individually. Why, then, should chemists bother to consider what would seem to be a purely imaginary realm? After all, isn't chemistry reputed to be an experimental science, based on direct observation? And aren't there enough mysteries to be solved right here in the macroscopic† world—the world of direct observation—to keep an army of

† *Macroscopic* comes from the Greek words for "large" and "to view," and hence, literally, applies to visibly large objects.

chemists busy for a thousand years? The answer to these questions is that chemists are not satisfied with merely observing nature; they want to understand and explain their observations. To do this the creation of a submicroscopic world is indispensable.

Suppose, for example, that we are asked to explain why the pressure exerted by the air in an automobile tire increases after the automobile is set in motion. The explanation we might offer is that heat is generated as a result of the friction between the moving tire and the road, that the temperature of the air inside the tire consequently rises, and that this increase in temperature is responsible for the increase in pressure. If this statement were not sufficient, we might add that air is a gas, and that the pressure exerted by air, or by any other gas, has always been found to increase whenever the gas is heated in a container of fixed size. So it should come as no surprise that the pressure in the tire increases also, since this is merely another example of the same well-known phenomenon.

Many people might accept this as a satisfactory explanation. But reflection will show that this kind of explanation raises a new and even larger question: Why does the pressure exerted by *any* gas in a fixed container always increase when the temperature rises? This is a valid question, and the phenomenon cannot be fully explained until an answer can be given.

Since any gas will display the pressure increase, and since the phenomenon is independent of the material used for the container, it seems logical to suppose that the pressure increase is connected with some feature of the internal structure of gases rather than with some property of the container. In trying to find the answer, one would therefore examine various gases by all sorts of techniques in the hope of perceiving such a structural feature. When this is done, however, no direct perceptual evidence is found.

If the examination has been thorough and imaginative, we may decide that the failure to obtain direct perceptual evidence is not due to lack of skill in observation, but rather to the limitations set by the human senses. In other words, the structural features we have been hoping to observe are too small to be perceived. When faced with such a conclusion, the physical scientist will usually resort to explanation by means of a submicroscopic *model.*

To the physical scientist a model is one of two things: it may be a picture he draws of the world as it would appear if our senses were superhuman (unlimited in quality or in range, large or small) or the model may be purely mathematical, and so abstract and general in scope that a pictorialization of it is not possible.

The picture models are relatively easy to understand and use, for all of us have been conditioned since childhood to accept as "real" a great many things that we have seen only in the form of pictures or toys. Fortunately, most of chemistry can be explained by means of picture models, and we shall use them often. However, the picture models frequently prove to be inadequate because nature is more complex than any model that is simple enough to be visualized. Moreover, a picture model, like a photograph, focuses attention on a particular limited aspect of nature and may not be as general as we would like. A mathematical model, which consists generally of an equation or a set of related equations, is not subject to limitations of this kind. Furthermore, manipulation of the equations frequently brings to light new relationships that were not envisaged by the inventors of the equations. A few mathematical models will be mentioned later, but only when the picture models seem to be inadequate.

Explanation by means of submicroscopic models

Explanation in terms of submicroscopic models usually involves the following steps:

1. The event or process to be explained is clearly stated, usually in terms of the experimental procedure which led to its observation.
2. Existing models are examined to determine whether or not they can offer an explanation.
3. If existing models are inadequate, a model is *invented* to explain the phenomenon.
4. The invented model is examined for logical consequences that can be confirmed or denied by further experiment.

An example of a *picture model* will help to make this concept clearer. The phenomenon to be explained is this: A gas confined in a container exerts a pressure on the walls of the container. To explain this observation, let us propose the following model:

1. The gas consists of a large number of tiny particles.
2. There is enough empty space between the particles so that the particles are free to move.
3. The particles are constantly in motion.

Let us see whether this model can account for the observation by picturing in our minds what might be happening. Because there are a great many particles in constant motion, the walls of the container are constantly being bombarded by particles hitting them. Every time such a bombardment occurs, the particle exerts a force or push on the wall while being deflected back into the inside of the container. If the number of such bombardments of the walls by individual particles in unit time were very large, the impression received by a human observer would be that of a steady pressure.

The same model, if made only slightly more elaborate, is capable of explaining why the pressure exerted by a confined gas increases with the temperature. In addition to the three assumptions made above, let us suppose also:

4. The speed at which the particles are moving increases with increasing temperature.

If this assumption were correct, the particles, moving faster at the higher temperature, would collide more often with the walls of the container; and whenever they did collide, they would exert a greater force. Thus, the impression received by an observer would be that of an increased pressure.

A model in which matter is ultimately composed of tiny particles that are in constant motion is quite different from the impression given us by our senses. According to our senses, a gas seems uniform and continuous. Therefore, such a model would hardly be taken seriously by many scientists unless it were also able to explain a great many other phenomena. As it turns out, the same basic model, with varying degrees of elaboration, can explain an enormous number of seemingly unrelated phenomena. It can explain such things as why heat flows from a region of higher temperature to one of lower temperature, how chemical reactions take place, or how vitally needed foodstuffs get through the walls surrounding living cells.

Once the basic model has been conceived, its elaboration becomes of first order of importance. Some phenomena can be explained only if the submicroscopic particles envisaged in the basic model have a certain size, or weight, or composition, or shape, or if some of the particles travel at certain speeds, or if they spin, or if they are tiny magnets. As the model becomes more and more elaborate, the number of macroscopic phenomena that can be explained grows larger and larger, until eventually we arrive at a picture of matter that will explain *all* of the behavior of *all* matter. This, of course, is the ultimate goal of the physical sciences.

"Attitude," not "method"

The continual switching back and forth between the macroscopic world of observation and the submicroscopic world of theory has been a highly fruitful source of scientific progress. It is indispensable to scientific reasoning. If there were such a thing as a tried-and-true scientific method, it would have to be part of it.

However, in our opinion there is no such thing as a unique "scientific method." Precisely because scientists are on the frontiers of their fields, they do not know where they are going, and therefore cannot set up sure-fire ways of getting there. What is common to all good scientists is better described as the *scientific attitude.* This attitude combines a burning curiosity with an open mind and a genuine interest in advancing knowledge. The scientist must have the courage to uphold his scientific convictions, but also the wit to abandon his theories when they are found wanting. He must be a skeptic, critical of his own work and of that of his fellow scientists. He must recognize that books and scientific journals are full of data and interpretations that can be accepted only as tentative. Experiments and conclusions must be checked and rechecked before they can be accepted as reliable. Even data and interpretations of long standing are sometimes found to be in need of revision and reinterpretation.

Groping for knowledge is a highly personal procedure; it takes many forms because each scientist does his research in the manner best suited to his particular temperament. Some, who have a strongly developed scholarly temperament, sort out their ideas and plan their experiments very carefully before they take a single step in the laboratory; others, who are less patient but have an equally clear idea of the problem, prefer to attack it directly in the laboratory. Both methods have their important advantages: the careful planner is more likely to solve his problem without wasteful motion; and the vigorous experimenter is more likely to open up new avenues for further research. Most scientists display both types of behavior at various times in their career.

Research is often most exciting when the results of experiments are totally at variance with what past experience might have led one to expect. The scientist is happiest when his research has created a new problem which demands further research.

The scope of chemistry

Chemistry has been defined as the science that deals with the structure of matter and the transformations it undergoes. This is such a broad definition that it might be helpful to expand a bit upon this definition and consider more specifically what constitutes the field of chemistry. This has become an increasingly difficult task because chemists have so broadened the scope of their interests that they now include much of what used to be the province of the biologist and the physicist. At one end of the spectrum of interests we find the chemical physicists and physical chemists, who are concerned largely with fundamental questions about the structure and properties of atoms and molecules and with formulating the mathematical theories that explain those properties. At the other end of the spectrum we find the biochemists and the molecular biologists, whose concern is with the very complicated molecules and chemical reactions found in living organisms. Somewhere in between the physical and biological ends of the spectrum are to be found chemists interested in inorganic or organic chemistry. The former is concerned with the chemistry of metals and non-metals other than carbon. The latter specializes in the chemistry of carbon compounds of which the complex organic world is largely composed. In addition, there are specialized areas in which these disciplines overlap and merge. In what follows in this book we shall concentrate on general principles that underly *all* of chemistry.

Suggestions for further reading

Armitage, F. P., *History of Chemistry* (London: Longmans, Green & Company Ltd., 1920).

Beveridge, W. I. B., *The Art of Scientific Investigation* (New York: Random House, Inc., 1957).

Brown, J. C., *History of Chemistry* (London: J. & A. Churchill, 1920).

Brown, M. G., *Critical Readings in Chemistry* (Boston: Houghton Mifflin Company, 1969).

Einstein, A., *Essays in Science* (New York: Philosophical Library, 1934).

Farber, E., *The Evolution of Chemistry* (New York: The Ronald Press Company, 1952).

Ferchl, F., and Süssenguth, A., *A Pictorial History of Chemistry* (London: William Heinemann Ltd., 1939).

Holton, G., and Roller, D. H. D., *Foundations of Modern Physical Science* (Reading, Mass.: Addison-Wesley Publishing Co., Inc., 1958).

Inde, A. J., *The Development of Modern Chemistry* (New York: Harper & Row, Publishers, 1964).

Jaffe, B., *Crucibles: The Story of Chemistry* (Greenwich, Conn.: Premier Books, Fawcett Publications, Inc., 1962).

Platt, J. R., *The Excitement of Science* (Boston: Houghton Mifflin Company, 1962).

Schneer, C. J., *The Search for Order* (New York: Harper & Row, Publishers, 1960).

Smith, H. M., *Torchbearers of Chemistry* (New York: Academic Press Inc., 1949).

Theobald, D. W., *An Introduction to the Philosophy of Science* (London: Methuen & Co. Ltd., 1968).

Tilden, W. A., *Famous Chemists* (New York: E. P. Dutton & Co., Inc., 1921).

Vavoulis, A., and Colver, A. W., *Science and Society, Selected Essays* (San Francisco: Holden-Day, Inc., 1966).

Waddington, C. H., *The Scientific Attitude,* (rev. ed., Baltimore: Penguin Books, Inc., 1948).

Weeks, M. E., and Leicester, H. M., *Discovery of the Elements,* 7th Ed. (Easton, Pennsylvania: Chemical Education Publishing Co., 1968).

Suggestions for supplementary textbooks

Beiser, A., and Krauskopf, K. B., *Introduction to Physics and Chemistry,* 2nd Ed. (New York: McGraw-Hill Book Company, 1969).

Quagliano, J., and Vallarino, L., *Chemistry,* 3rd Ed. (Englewood Cliffs, N.J.: Prentice-Hall, Inc., 1969).

Sienko, M., and Plane, R., *Chemistry,* 3rd Ed. (New York: McGraw-Hill Book Company, 1966).

Sisler, H. H., Vanderwerf, C. A., and Davidson, A. W., *College Chemistry,* 3rd Ed. (New York: The Macmillan Company, 1967).

Questions

1. Study Appendix A: *Some Scientific Units.*

2. Study Appendix B: *Exponential Notation.*

My advice to those who wish
to learn the art of scientific prophecy
is not to rely on abstract reason,
but to decipher the secret language
of Nature from Nature's documents,
the facts of experience.

MAX BORN (1882–1968)

Two

MIXTURES AND PURE SUBSTANCES

THE MATTER we encounter in our every day experience is almost always a complex mixture. The things that are most familiar to us, and which we might expect to be easy to classify and study, are of such complexity as to make the analytical chemist shudder. The air we breathe is a mixture of at least five substances; gasoline is a complex mixture of more than a dozen substances; Portland cement consists of five or more substances; steel, though consisting largely of iron, owes its properties to the admixture of a number of other substances during its manufacture. Virtually all living or once-living forms of matter are complicated almost beyond description; it would be a hopeless job to count all the different substances present in the human body.

It is quite unusual for the matter that we encounter in ordinary experience to be "pure," that is, to consist of only a single substance. Distilled water is a pure substance; so is the copper that we use in electrical wiring; the granulated sugar that we add to our coffee; baking soda, some mothflakes, and certain medicinal drugs. But one is hard put to extend this list much further. Even such a pure-looking substance as ordinary table salt consists not only of sodium chloride crystals, but also of small amounts of additives that delay clumping.

Because of the great complexity of ordinary matter, chemistry as a science had rather a slow start. It was only after the necessity for studying the behavior of "unmixed" substances had been recognized, and methods for separating the ordinarily encountered mixtures into their components had been perfected, that the development of chemistry became rapid. The wonderfully simple and straightforward laws of chemical combination and composition were discovered only after chemists began to deal with relatively pure materials.

This procedure—the separation of complex mixtures into their components which are then studied individually—illustrates a general strategy of the laboratory sciences. Complex, seemingly insoluble problems can often be solved by breaking them down into a number of smaller problems, each amenable to solution.

Physical properties of matter

The concept of a pure substance is so important that it deserves careful consideration. Two examples in which the complexity of matter is evident even on casual observation are a T-bone steak and, on a less digestible level, concrete pavement. In the former, one can easily discern regions of bone, fat, and meat, while in the latter, one can see sand grains, gravel, and hardened cement. In both examples, we know just from their appearance that we are dealing with mixtures of substances. Matter with these characteristics is referred to as *heterogeneous*.†

There are, however, a good many materials that are perfectly uniform throughout, that is, any one portion is just like every other portion. Nevertheless, in many cases, these materials consist of more than one substance. Examples that have been mentioned are air, gasoline, and steel. Materials that are uniform throughout are referred to as *homogeneous*.‡ When a homogeneous material consists of more than one substance, it is called a *solution*. From the examples that have been given, it is clear that a solution may be gaseous, liquid, or solid. Figure 2-1 shows the subdivisions of different types of matter.

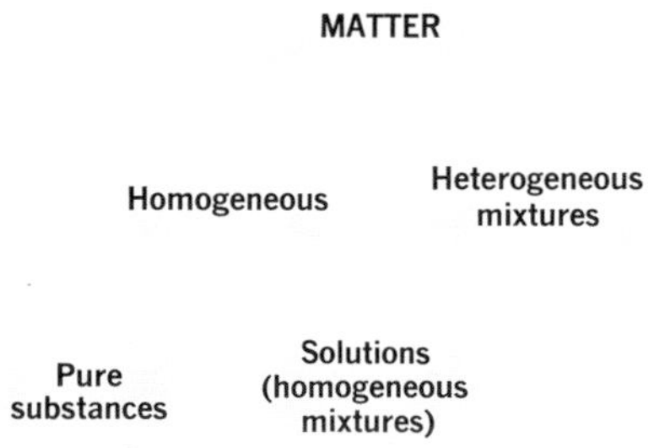

MATTER

Homogeneous

Heterogeneous mixtures

Pure substances

Solutions (homogeneous mixtures)

Fig. 2-1

Types of matter.

† A term derived from a Greek root meaning "of other kinds."

‡ A term derived from a Greek root meaning "of the same kind or nature."

A difficult problem facing the chemist is to distinguish between solutions and pure substances, since both are perfectly uniform even upon microscopic examination. He solves the problem by subjecting the homogeneous sample to one or several of the procedures that are capable of separating mixtures into their pure components. If these operations lead to the separation of the sample into a number of substances with different properties, then he reasons that the original sample must have been a solution. On the other hand, if the properties of the sample remain unchanged, then he is reasonably sure that he is dealing with a pure substance.

What are these properties that may or may not be changing? The most obvious ones are those that can be noted with our unaided senses, such as the color of the material, or its odor, or perhaps even its taste. During the nineteenth century it was not unusual for a chemist to sample the taste of something he had just prepared—without swallowing it, of course. But the unaided senses are not sensitive enough to detect small differences in properties. Moreover, such tests are subjective, and the results vary with the individual making the test. It is therefore more conclusive to examine other properties that can be measured objectively with appropriate instruments. Some of those properties are listed and defined below. (See Appendix A for a discussion of scientific units and of temperature scales.)

1. *The density* is the weight per unit volume. The density is usually expressed in grams per cubic centimeter (g/cc).
2. *The melting point* (m.p.) of a solid is the temperature at which the solid is converted to a liquid.
3. *The boiling point* (b.p.) of a liquid is the temperature at which the liquid is converted rapidly into vapor, the conversion being accompanied by the formation of bubbles. The boiling point varies with the prevailing pressure, but is normally reported at a pressure of one atmosphere (atm).
4. *The refractive index* is related to the speed at which light travels in the given sample of matter. In practice, the refractive index is measured easily and accurately by means of the angles at which the light enters and leaves the substance.
5. *The conductivity for electricity* is a measure of the ease with which an electric current can pass through the substance. Copper is an

excellent conductor, but glass is a poor conductor or good insulator.

6. *The viscosity* of a liquid or gas is a measure of the ease with which the liquid or gas will flow. For example, water will flow much more readily than motor oil.
7. *The absorption spectrum of a substance* is a catalogue of the wavelengths at which the substance absorbs visible, infrared, and ultraviolet light.

The density, refractive index, conductivity for electricity, and viscosity all vary with the temperature and must therefore be measured after the sample has been brought to some definite temperature, often chosen to be 25°C (77°F).

The operational definition of a pure substance

Let us suppose that we want a sample of really pure water. In most communities, the tap water is pure only in the medical sense, that is, it is free from harmful or poisonous ingredients. The tap water is not pure in the chemical sense since it usually contains small amounts of dissolved minerals and gases, which we shall call *impurities*.

Our job, then, is to obtain chemically pure water from tap water. There are a number of procedures that one might use, some of which will be described later in this chapter. A convenient method is to cool the water and to allow about half of it to freeze. The impurities tend to remain in the liquid, and the solid ice that freezes out is likely to be quite pure. The ice may then be separated from the liquid, placed in a separate container, and allowed to melt. This process may be repeated as often as necessary.

Another method of purification, which we shall discuss later, is distillation. The distilled water is more nearly pure than the original undistilled sample.

After each purification step, the purified sample of water is examined and several of its physical properties are determined. There will probably be a difference between the properties of the tap water and those of the water sample obtained in the first purification step, indicating that some impurities have been removed. As the purified water is subjected to further purification, the differences in properties of the doubly, triply, and so on, purified water become progressively smaller, and eventually a sample is obtained whose properties are not changed on further purification. Such a sample is then accepted as chemically pure.

Some of the properties of pure water are as follows: pure water boils at 100°C (at one atmosphere pressure); it has a refractive index of 1.333 and a density of 0.997 g/cc at 25°C; pure ice melts at 0°C. These are the properties of *all* samples of pure water, regardless of their origin, the nature of the impurities they contained before purification, and the location of the laboratory where the purification took place. Hence we conclude that these properties are characteristic of the substance, water. If in the future we encounter another liquid with boiling point 100°C, density 0.997, refractive index 1.333, and identical in all other observable respects with an authentic sample of pure water, then we assert that that liquid must also be pure water.

Reflection will show that the method of recognizing a pure substance by its properties is exactly the method we use to recognize objects or persons in everyday life. Thus, we recognize a friend by the shape of his head, the color of his hair, or the tone of his voice; and in a similar manner we recognize a pure substance by its boiling point, density, or other properties. There is one important difference, however. While the friend may alter his appearance at will, the physical properties of a pure substance under a given set of conditions are characteristic of the material and are inseparable from it. If Russ gets a suntan and grows a beard, he is still Russ, but if a liquid boils at 80°C (at one atmosphere pressure) it cannot be water.

In short, a given substance is simply the sum of its properties, and can thus be recognized by these properties under any circumstances. When you ask "What is iron?" the best answer is that iron is a substance that melts at 1535°C, boils at 3000°C and one atmosphere pressure, has a density of 7.86 at 20°C, reacts with oxygen, and so on. Since the complete set of properties is unique for any given pure substance, it is obvious that if two samples of matter have identical sets of properties they are samples of the same pure substance.

We have defined a pure substance by means of experimental operations: the substance is accepted as pure when attempts at further purification produce no further changes in its properties. Such a definition is called an *operational definition.* Operational definitions are essential to the success of any experimental science. An operational definition is one that prescribes a specific experiment or measurement to define a concept or situation. We say, "Do thus and so, and what you will observe will give the answer to your question."

Operational definitions are fundamentally different from theoretical definitions, in which the given concept is defined in terms of a particular theoretical model. Thus, for the concept *pure substance,* there is an alternative definition based on the theory that matter is ultimately composed of submicroscopic particles called *molecules.* In terms of that theory, a pure substance is defined as a sample of matter in which all the molecules are alike.

The states of matter

The preceding classification of matter on the basis of its purity complements the common classification of matter on the basis of its *state,* which may be solid, liquid, or gaseous. Water can exist in all three states: as ice, as liquid water, and as steam. Water is one of the few substances with which the average person is familiar in all three of its states. The reason is simple: the temperature range over which transitions among the various states take place is quite narrow (100° on the centigrade scale) and the melting point and boiling point are not too far from room temperature.

Most other common substances are familiar to us in only one state, because the transition temperatures are far from room temperature. An example is the substance iron, familiar only as a solid. However, at a temperature of 1535°C iron becomes liquid, and at 3000°C and one atmosphere pressure it is converted to a gas (a vapor). On the other hand, oxygen is best known as a gas. However, at a temperature of −183°C and one atmosphere pressure it becomes a liquid, and with further cooling it solidifies at −218.4°C.

The three states of matter can be defined operationally in terms of their macroscopic characteristics (we will defer until later a consideration of how they are defined theoretically in terms of submicroscopic particles), as follows:

The *solid state* is characterized by

1. high density, usually in the range from 1 to 10 grams per cubic centimeter;
2. rigid shape;
3. slight compressibility;
4. slight expansion at atmospheric pressure as the temperature is is raised (usually less than 0.01% per degree centigrade).

The *liquid state* is characterized by

1. high density, but usually somewhat less than that of the corresponding solid;
2. lack of definite shape (liquids conform to the shape of the containing vessel);
3. slight compressibility, but usually somewhat greater than that of the corresponding solid;
4. slight expansion at atmospheric pressure as the temperature is raised (but somewhat greater than that of the corresponding solid, usually about 0.1% per degree centigrade).

The *gaseous state* is characterized by

1. low density, of the order of 0.0001 to 0.004 grams per cubic centimeter at one atmosphere;
2. indefinite shape and volume (gases fill completely any container into which they are placed);
3. high compressibility;
4. considerable expansion (about 0.3% per degree centigrade) when heated at constant pressure.

Isolation and purification of substances

Since the world is made up largely of mixtures of varying complexity, and since chemical reactions become intelligible only when we deal with pure substances, it is of the utmost importance to have reliable methods by which mixtures can be separated into their constituent substances.

If the mixture is not homogeneous, then it is sometimes possible to separate the various homogeneous components by mechanical means. For example, gold is separated in this way from the accompanying gravel. Usually such mechanical means are not very successful, however, and other techniques must be used.

In the following pages, we shall discuss several techniques that have been developed for the separation of mixtures and the isolation of the pure components. We shall describe each technique as it is used in the laboratory with fairly small samples, but the reader should realize that these operations are also carried out commercially with carload quantities of materials in the chemical industry, using large-scale modifications of the equipment.

1. DISTILLATION

Distillation depends on the fact that the components of a liquid solution† usually differ in their volatility,‡ that is, in the ease with which they can be vaporized. When a solution is heated to boiling, the vapor that boils away contains a higher concentration of the more volatile component than the original liquid. If this vapor is now collected separately and condensed back to the liquid state, the new liquid condensate obviously will be richer in the more volatile component. The process of boiling followed by condensation may be repeated many times, until eventually the condensate is so highly enriched in the more volatile of the two components that it may be regarded as a pure sample of the more volatile component.

To see how purification by distillation works, let us consider a liquid solution containing equal weights of water and acetone. Acetone is an organic liquid that is often used in chemical operations, especially as a solvent for other substances. As we know, water boils at 100 °C. Acetone is more volatile than water, and pure acetone boils at 56°C. The 50–50 mixture initially boils at an intermediate temperature, which has been measured and found to be 66°C. This boiling point gradually rises, as the more volatile acetone vaporizes, until it reaches that of pure water.

The vapor obtained from the 50–50 mixture initially contains 92%

† See page 14 for a working definition of "solution."

‡ *Volatile* is from the Latin "to fly"; as an English word it also means "giddy, lighthearted."

of acetone—an enormous enrichment. If the vapor is condensed, the resultant liquid of course has the same composition, namely, 92% acetone. If 92% liquid is boiled and its vapor condensed, the new liquid is further enriched in acetone, containing 97% acetone. The process of boiling-followed-by-condensation may be repeated as often as necessary until acetone of any desired degree of purity is obtained.

The apparatus for carrying out a simple distillation is shown in Fig. 2-2. The liquid is brought to a gentle boil in the boiling flask *A*, and the vapor rises into the condenser tube *B*, the walls of which are cooled, normally with tap water. Since the temperature of the condenser is below the boiling point, the vapor condenses to a liquid and is collected in the receiving flask, *C*. Since the vapor is always richer in the more volatile component, the liquid remaining in the boiling flask becomes gradually richer in the less volatile component as more of the more volatile component boils away. Eventually, when virtually all of the more volatile component has been distilled, the liquid remaining is a nearly pure sample of the less volatile component. The solution has thus been separated into its pure components.

Devices are available in which the process of boiling-followed-by-condensation is carried out time after time in a single apparatus. One such apparatus is shown in Fig. 2-3. It differs from the simple apparatus of Fig. 2-2 by the inclusion of a so-called fractionating column, *D*, in which this repetitive process occurs.

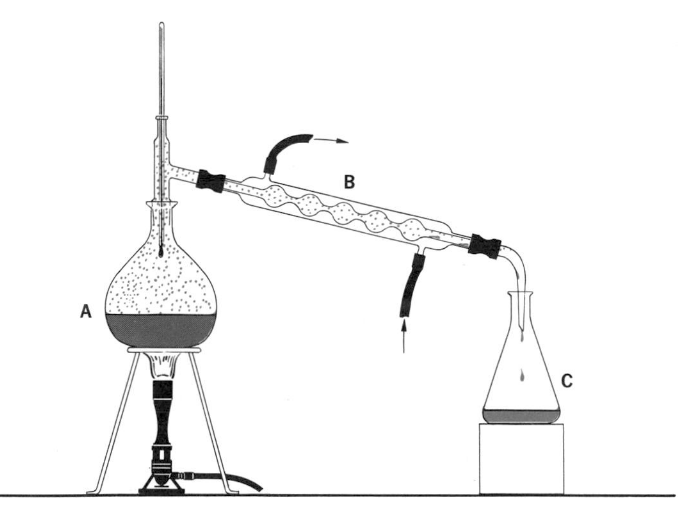

Fig. 2-2
A simple distillation.

2. RECRYSTALLIZATION

Recrystallization is an excellent method for the purification of slightly impure solid substances. The method depends on the fact that the solubility of most solid substances in a given liquid increases markedly with increasing temperature. The slightly impure solid is dissolved in a suitable quantity of a liquid (perhaps water or alcohol) which is kept at or just below its boiling temperature. When most or all of the solid has been dissolved, the hot solution is filtered, that is, the solution is allowed to drip through a filter paper or a porous glass disk which permits the liquid to pass but retains specks of dust or other solid impurities. The hot filtered solution is then set aside to cool. As the temperature drops, the solubility of the solid decreases, and the solid starts to crystallize. The crystals are usually quite pure since the impurities tend to remain in the solution. The crystals are then separated from the remaining liquid by filtration. The purity of the crystals is usually measured by the sharpness of their melting point, that is, by the narrowness of the temperature range over which melting occurs. If a sample is perfectly pure, all of it will melt at the same temperature.

Some laboratory apparatus used in recrystallization is shown in Fig. 2-4.

Fig. 2-3
The fractionating column, D, is packed with glass rings which present the upward-moving vapor with a large surface area on which condensation and boiling can take place repeatedly.

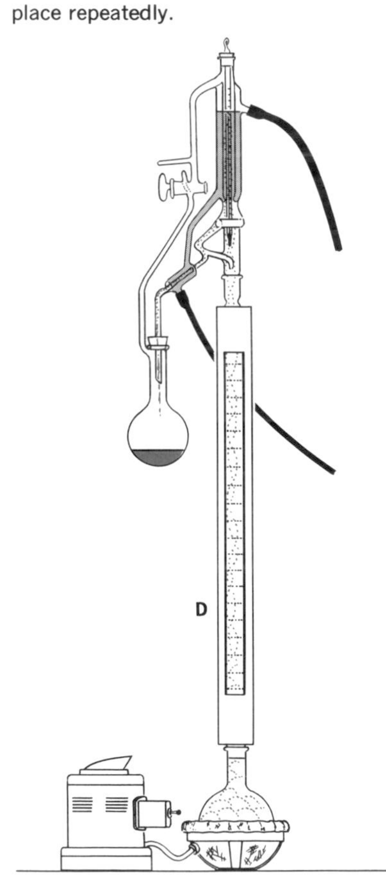

3. SOLVENT EXTRACTION

Solvent extraction is often used in order to separate a specific substance from a liquid solution containing several substances. The method works as follows: the original solution is placed into a *separatory funnel* (shown in Fig. 2-5), and a second liquid, which we shall call the *extracting solvent,* is added. The extracting solvent must possess two essential characteristics: it must be virtually insoluble in the initial solution, and it must be a good solvent only for the substance to be extracted.

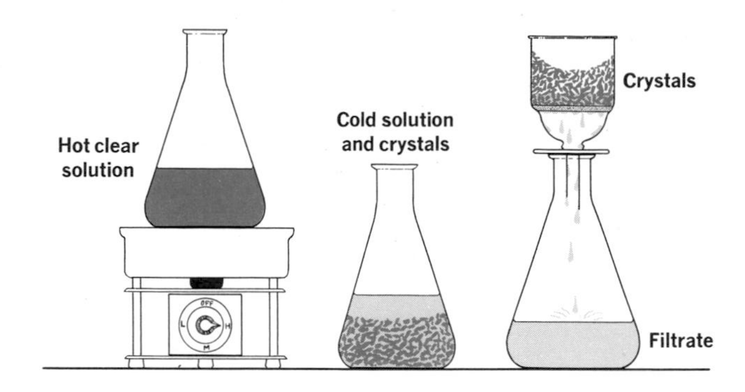

Fig. 2-4
Steps in recrystallization.

For example, let us suppose that we wish to separate iodine from a solution containing water, iodine, and sugar. We place the solution into a separatory funnel and add a roughly equal volume of the liquid *carbon tetrachloride,* which possesses the correct characteristics of an extracting solvent for iodine. Carbon tetrachloride does not dissolve in water, and neither water nor sugar is soluble in carbon tetrachloride. But iodine is a hundred times more soluble in carbon tetrachloride than in water. That is, if the separatory funnel is shaken so that the two liquids are brought into close contact, a hundred parts of iodine will pass into the carbon tetrachloride for every one part remaining with the sugar in the water solution. The carbon tetrachloride, which is more dense than water, is then drawn off through the stopcock at the bottom

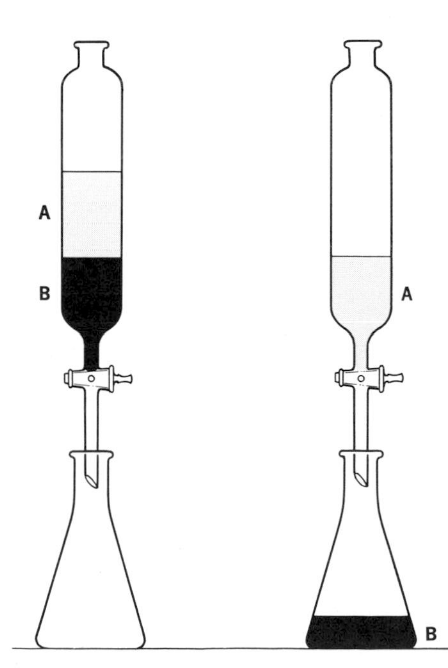

Fig. 2-5
Separation of two immiscible solutions A and B by means of the separatory funnel. The denser liquid (B) is drained off through the stopcock.

of the separatory funnel, leaving behind the sugar and water. When the carbon tetrachloride is evaporated, the solid residue is quite pure iodine.

4. CHROMATOGRAPHY

Chromatography makes use of the fact that the surfaces of solids are able to attract other substances and hold them more or less tightly. This process is known as *adsorption*. For ordinary solids, the amount of other substances that can be adsorbed in this way is rather small, but when the solid is given a large surface area by being ground to a very fine particle size, it can adsorb appreciable amounts. The tenacity with which different substances are held by a given adsorbing surface can vary widely, and when it does vary, the substances can be separated. The method works as follows:

Let us suppose that we wish to separate a mixture of two substances that differ greatly in the firmness with which they are adsorbed by a given solid. After the mixture has been adsorbed, the surface carrying the two substances is washed with a liquid that is capable of dissolving both substances but not the adsorbing surface. The substance that is held less firmly washes off relatively easily and can therefore be separated from the other substance, which washes off less easily. In the laboratory, this operation is carried out most easily by packing a straight glass tube with the adsorbing solid. The mixture of two or more substances to be separated is introduced at the top, and then is gradually moved down the column by repeated addition of a suitable liquid. This liquid carries each component of the mixture down the column at a rate that depends on how firmly the component is adsorbed. Those components that are adsorbed most firmly move most slowly. The differences in the rates of movement result in the separation of the mixture into distinct bands, each representing one component of the mixture, as is shown in Fig. 2-6. Originally, this method was used to separate mixtures of colored substances whose passage down the tube could be followed visually,† but methods are now available for applying the method to colorless substances.

The technique of chromatography has been extended to include the separation of a wide variety of substances in the vapor phase. In the

† Hence the name *chromatography*, from a Greek word meaning "color writing."

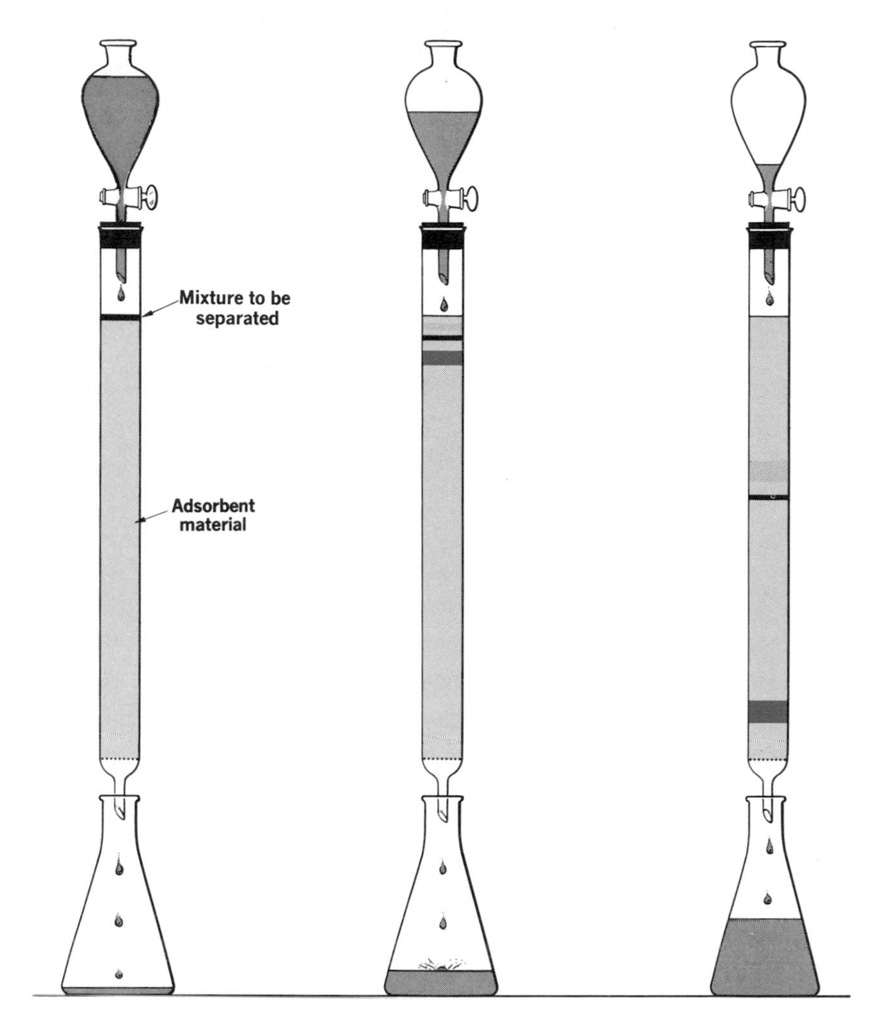

Fig. 2-6
Separation by chromatography.

simplest case, the technique is quite similar to that described above: A mixture of substances in the vapor phase is separated by differential adsorption on a solid material, using a gas such as helium as the *carrier fluid.* The components of the mixture become separated as they pass through the column and emerge one-by-one, just as in the case of liquid chromatography.

A modification of this method which allows for even broader application of the chromatographic technique, is known as gas-liquid partition chromatography. In this method, the chromatographic column is filled with chips of a chemically inert solid, such as firebrick, that have been coated with a viscous liquid, such as vaseline. The gases or vapors, instead of being adsorbed on the surface of the solid particles, are dissolved in the viscous liquid around each particle. Since each substance in the gas or vapor has a characteristically different solubility, a separation of the mixture is possible.

The application of chromatography has resulted in the separation and analysis of extremely complex mixtures of closely related substances. In the petroleum industry, for example, the analysis of mixtures containing hundreds of substances, some present as only a tiny fraction of 1%, has now become a routine matter. The method has also greatly increased the effectiveness of research in such areas as photochemistry and radiation chemistry, where the reaction products are often complex mixtures.

By means of these processes and others, the complex mixtures ordinarily encountered in nature can be separated. Since the fundamental nature of matter is revealed most clearly from the behavior of *pure* substances, the methods of separation are of enormous importance. Without the ready availability of such pure substances, the development of modern chemistry would have been impossible.

Suggestions for further reading

Astin, A. V., "Standards of Measurement," *Scientific American,* June 1968, p. 50.

Fullman, R. L., "The Growth of Crystals," *Scientific American,* March 1955, p. 74.

Gaucher, G. M., "An Introduction to Chromatography," *Journal of Chemical Education,* November 1969, p. 729.

Holden, A., and Singer, P., *Crystals and Crystal Growing* (Garden City, N.Y.: Anchor Books, Doubleday & Company, Inc., 1960).

Keller, R. A., "Gas Chromatography," *Scientific American,* October 1961, p. 58.

Mason, B. J., "The Growth of Snow Crystals," *Scientific American,* January 1961, p. 120.

Pfann, W. G., "Zone Refining," *Scientific American,* December 1967, p. 62.

Rabinowicz, E., "Polishing," *Scientific American,* June 1968, p. 91.

Snyder, A. E., "Desalting Water by Freezing," *Scientific American,* December 1962, p. 41.

Questions

1. Go to the library and examine the index of a handbook of chemistry and physics. Find ten entries in the index that represent definite properties of pure substances.

2. Using a handbook of chemistry and physics, look up the melting points and/or boiling points of the following inorganic substances:

(*a*) Sodium
(*b*) Sodium iodide
(*c*) Hydrogen
(*d*) Water (hydrogen oxide)
(*e*) Lead
(*f*) Lead chloride

3. One ounce of finely granulated pure white sugar (sucrose) and one ounce of finely granulated pure white salt (sodium chloride) are placed in a bottle. The bottle is stoppered tightly and the mixture is shaken until the two solids are intimately mixed. Will the resulting mixture be "homogeneous" in the sense in which that term is used by chemists? Explain carefully.

4. A manufacturer of aspirin tablets advertises "Our tablets are 100% pure aspirin." He probably means that his tablest are pressed from chemically pure aspirin, without an inert filler. How would you check his claim?

5. A given clear colorless liquid is known to be a solution of either sugar or salt in water. How would you find out which it is?

6. A chemist made up two solutions, one of acetone (a clear colorless liquid) in water, the other of salt in water. He placed the two solutions into separate bottles, but forgot to label them. Suggest a simple experiment to find out which is which.

7. Give examples of a solution of:
(*a*) Two solids in one another.
(*b*) Two or more gases.
(*c*) One common food item in another.

8. Is the following material homogeneous or heterogeneous?
(a) Liquid shoe polish.
(b) Waxy shoe polish.
(c) Glass used in eye-glasses.
(d) Clear vinegar.
(e) Paper used for stationery.
(f) House paint.

9. Which of the methods described in the chapter would be suitable for separating the following mixtures into their pure components?
(a) Crude oil.
(b) Impure aspirin.
(c) A solution of silver (B.P.961°C) in mercury (B.P.357°C).
(d) Purple ink.

Seek simplicity, and distrust it.

ALFRED NORTH WHITEHEAD (1861–1947)

Three

ELEMENTS AND COMPOUNDS

IN THE preceding chapter we discussed the nature and preparation of pure substances and emphasized the importance of dealing with truly homogeneous substances. In this chapter we shall discover that there are two classes of pure substances, *elements* and *compounds*. In order to define these classes operationally it will be necessary first to consider the macroscopic characteristics of chemical change.

The macroscopic characteristics of chemical change

The rusting of iron is a familiar example of chemical change. The strong, ductile metal is slowly transformed into a brittle, easily crumbled, red powder of little mechanical strength. This red powder is a new substance in no way like the parent metal. For example, the magnetic property commonly associated with iron no longer exists in this new material, its density is less, and it is no longer a good conductor for electricity.

The process of combustion in which coal, wood, or oil is consumed is another dramatic chemical change. The products of combustion of wood in no way resemble the original wood, these being largely a gas (carbon dioxide), water, and ash. The conversion of the food we eat to living tissue, the growth of plants, the souring of milk, and the decay of vegetation are all examples of chemical changes being carried on by living organisms.

The cooking of an egg, the conversion of high-boiling crude oils into gasoline, the production of metallic aluminum from the stony-looking mineral bauxite, the hardening of mortar, the evolution of gas that occurs when baking powder is moistened, these are all further examples of the literally millions of chemical changes known to chemists.

The characteristics which *all* chemical changes (also called *chemical reactions*) have in common are:

1. The substances that are present initially (the *reactants*) disappear.
2. One or several new substances (the *products*) appear as the reaction proceeds and the reactants disappear. The properties of the products are recognizably different from those of the reactants.
3. Energy in the form of heat, light, or electricity is released or absorbed† in the course of the chemical change.

The elements

Man has always shown a desire to provide simple explanations for complex events. The earliest writings and legends reveal a preoccupation with a uniform, underlying reality. Thus, the Greek philosopher, Empedocles, proclaimed in the fifth century B.C. that the universe is composed of a multitude of combinations of the four "elements": *earth, air, fire, and water.*

A century earlier, Thales, another Greek, had proposed the theory that there is only one element, water, and that the multiplicity of the universe consists simply of variations of this one substance. In the years that followed, various other substances were championed as being the ultimate building blocks of the universe, but none was universally accepted. All of the various elements that were espoused suffered one serious shortcoming: the experiments that the various theories suggested did not lead to the development of a fruitful science of chemistry.

† Literally, absorb means "to suck away," i.e., to cause to disappear. A substance *ab*sorbed by another disappears into it, is mingled with it, whereas, as noted earlier, a substance *ad*sorbs another onto its surface.

Robert Boyle: 1627–1691
Boyle may be called the founder of modern chemistry. He recognized the intrinsic value of chemistry, in addition to its role as an adjunct to medicine. He introduced rigorous experimental methods and rebelled against the obscurantism of the alchemists. He was the first to distinguish between mixtures and compounds, and his definition of an element is still valid today. [Photograph courtesy Brown Brothers.]

It was not until 1661 that the British scientist, Robert Boyle, in his book *The Skeptical Chymist*, proposed a definition of an *element* that has withstood the test of time and experiment and, with slight modification, is still in use today. In Boyle's own words: "And to prevent mistakes, I must advertize You, that I now mean by Elements, . . . certain Primitive and Simple, or perfectly unmingled bodies; which not being made of any other bodies, or of one another, are the Ingredients of which all those call'd perfectly mixt Bodies are immediately compounded, and into which they are ultimately resolved."

In other words, Boyle is saying that there are certain substances (Boyle specifies no number) that can be thought of as being of ultimate simplicity, that is, they cannot be broken down into yet simpler substances; and it is these that, in various combinations, make up all of the complex substances (mixt Bodies) of the universe. Note that Boyle gives us a test, or experiment, the results of which can be used to determine whether or not a given substance is an element:

If the substance cannot be broken down into yet simpler substances by any means known to the chemist, it must be classified as an element.

The reader will note that Boyle's definition is an *operational* definition since a substance is classified as an element solely on the basis of experimental operations.

Thus, an element is any substance that will resist all known methods of decomposition. But in a growing science, there is always the possibility that new methods will be discovered which can break down some of the substances that had previously been thought of as elements. This has actually happened a number of times. For example, the substance potash was for a long time thought to be an element. However in 1807, by passing an electric current through moist potash, Sir Humphrey Davy demonstrated that potash is really a "mixt Body" consisting of familiar elements and of a new metallic substance which he called "potassium." The question remains whether, with potassium, we have at last arrived at an element, or whether potassium is yet another complex substance which is even more difficult to break down than the potash from which it was obtained. In the light of Boyle's definition, this question is answered by saying that all efforts at further decomposition have been resisted, and that potassium, therefore, meets all requirements and must be accepted as an element. However, according to Boyle's definition, we can never be sure that our list of the elements is final and complete.

Since Boyle's time, fortunately, independent theories have been developed, of which more will be said later, which give us criteria for making unequivocal statements about whether a given substance is an element or not. On that basis there are known today 88 naturally occurring elements and an ever-growing number of synthetic ones.† (See inside back cover for a list of the elements.)

According to Boyle's definition, an element must be a fairly stable substance. It must be at least stable enough to enable one to isolate it in the pure state, determine its properties, and see whether or not it can be broken down into yet simpler substances. Today, this definition is sometimes thought to be too rigid. On the one hand, some of the heavy synthetic elements are too unstable to satisfy these requirements. And on

† *Synthetic* is used here as a chemical term describing what is prepared artificially in the laboratory, as opposed to what occurs in nature. It is derived from a Greek word meaning "placed together."

the other hand, all of the substances that we call elements can be temporarily decomposed further under drastic conditions of high temperature or high voltage. Yet the discovery of such complications is of relatively recent date. For more than two centuries, during the critical early period in the development of modern chemistry, Boyle's definition permitted the rational organization of chemical knowledge and suggested fruitful experiments.

Compounds and mixtures

Let us now consider Boyle's "mixt Bodies" or *compounds,* as we shall call them.

A compound is a pure substance composed of two or more elements joined in chemical combination in a definite proportion by weight.

Thus a compound is *not* merely a mixture of the elements—a chemical reaction must occur between the elements. As a result of the reaction, the original properties of the elements have been lost and replaced by an entirely new set of properties that are characteristic of the compound. In reacting to form the compound, the elements have entered into a state of firm chemical combination; they are held together in the compound by chemical bonds that are so strong that the compound remains intact even when it is subjected to purification procedures, such as distillation or chromatography. It follows that compounds are pure substances, and we see that there are really two classes of pure substances: elements and compounds.

Let us consider a few examples of compounds in the light of this definition. Water consists of 11.2% by weight of hydrogen and 88.8% by weight of oxygen, in chemical combination. That is, hydrogen has reacted with oxygen to produce the new substance, water. Carbon dioxide is a compound consisting of 27.3% by weight of carbon and 72.7% by weight of oxygen, in chemical combination. Sodium chloride is a compound consisting of 39.3% sodium and 60.7% chlorine, in chemical combination. Sulfuric acid is a compound consisting of 32.70% sulfur, 65.26% oxygen, and 2.04% hydrogen, in chemical combination.

In each of these examples it must be emphasized that, if the percentages of the elements were significantly different from those cited, we would not have the compound in question. That is, if the percentages of the elements were different, the physical as well as the chemical properties would be different. By *chemical properties* we mean the types of reactions the substance will undergo, and the readiness with which these reactions take place.

The fact that the composition of a compound has a definite and fixed value gives us a criterion for distinguishing between compounds and mixtures. The percentage composition of a mixture is *not fixed* but can be varied arbitrarily. Thus the label on pancake syrup may state that it contains 15% maple syrup and 85% cane syrup, which might suggest that it is a compound of the corresponding elements. But the next brand we pick up could very well contain 20% maple and 80% cane syrup, and a third might contain 50% of each. In other words, we can vary the percentage composition in any way we like and still have a product that is correctly labeled pancake syrup.

The possibility for such arbitrary variation does not exist in the case of a pure compound. Sodium chloride always consists of 39.3% sodium and 60.7% chlorine, and any effort to produce sodium chloride of a different composition is futile. If one takes exactly 60.7 g of gaseous chlorine and 39.3 g of metallic sodium and allows them to combine, one obtains 100.0 g of the compound sodium chloride. Neither chlorine nor sodium is in excess, and all of each constituent is used up in the formation of the compound. Thus, if the constituent elements are allowed to react in the same ratio as they have in the final compound, there will be no uncombined elements left over. On the other hand, if instead of 39.3 g of sodium one uses 41.3 g and the same amount of chlorine as used before, chemical combination will again occur, and one will once more obtain 100.0 g of sodium chloride. However, 2.0 g of sodium will be left uncombined. The 2.0 g corresponds exactly to the difference between the amount demanded by the ratio of sodium to chlorine in the compound and the amounts actually used. Similarly, if excess chlorine were used, unreacted chlorine would be found at the conclusion of the experiment.†

† It is not possible, here, to have an excess of *both* elements. Thus, if we take 41.3 g of sodium and 62.7 g of chlorine (a 2 g excess of each), we find that the 2 g excess of chlorine reacts with the excess of sodium in the ratio required by the composition of the compound, leaving 0.7 g of unreacted sodium.

Behavior of this type is observed in all chemical reactions in which the elements unite to form only one compound. When more than one compound is formed, the situation is more complicated, but the composition of each of the compounds is still fixed and definite. Further discussion will be reserved for Ch. 4.

The formation of solutions

In most cases when a solution is formed, the substances involved in forming the solution are not permanently altered but can be recovered unchanged. An example is the dissolving of sugar in water. Evaporation of the water results in the recovery of unchanged sugar.

Sometimes, however, the formation of the solution is accompanied by a chemical reaction. When that happens, the constituents are profoundly altered and cannot be recovered in their original form by simple separation procedures. For example, when a metal is dissolved in an acid, hydrogen gas is liberated and heat is evolved. Evaporation of the solution leads to the recovery of a *compound* of the metal rather than of the original metal.

Intermediate cases exist in which the original substances can be recovered unchanged, but in which one may infer from the solution properties that some chemical interaction is taking place. For example, when sulfuric acid is added to water, there is a marked evolution of heat, suggesting that a chemical reaction takes place as these substances form a solution.

Physical change contrasted with chemical change

You may wonder how a given change is known to be a chemical change in the sense already discussed, rather than a physical change. In a physical change, a substance is heated, or cooled, or compressed, or otherwise exposed to some new external condition without losing its chemical identity. When the original external conditions are restored, the substance regains its original physical properties. On the other hand, in a chemical change the initial substance disappears and new substances with entirely different physical and chemical properties appear in its place.

Methods of bringing about chemical change

How can one *cause* a chemical reaction to occur? Frequently the simple act of bringing two substances together will result in their reacting with one another. (See Figs. 3-1 and 3-2 for examples of typical apparatus used in laboratories to produce chemical reaction.) Thus, if gaseous chlorine is brought into contact with metallic potassium, a violent *combination reaction* ensues, with the evolution of considerable energy. The product is potassium chloride, with none of the properties of the two reactants. Or if nitric oxide (a compound of nitrogen and oxygen) comes into contact with oxygen, a spontaneous reaction occurs and the new compound, nitrogen dioxide, is formed.

Fig. 3-1
Simple apparatus for carrying out a chemical reaction. One of the reactants is in the large flask, and a second reactant is added from the dropping funnel while the mixture is being stirred.

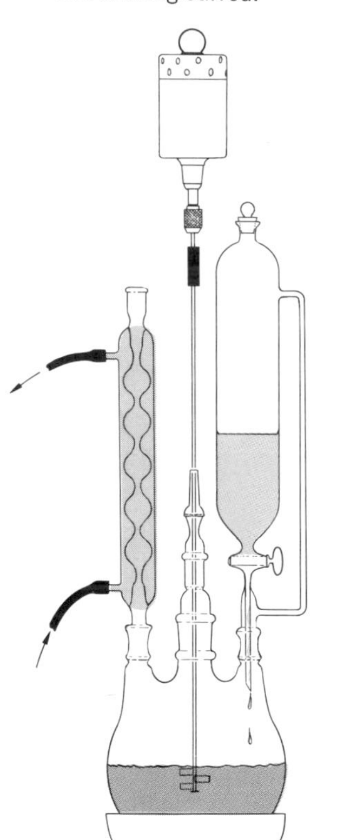

Single substances can also undergo chemical reactions. A common example is *decomposition,* in which a compound breaks down into simpler substances. Thus, when mercuric oxide (a compound of mercury and oxygen) is heated strongly, it decomposes into metallic mercury and gaseous oxygen. Another well-known type of reaction which involves only a single substance is *rearrangement,* in which the original substance disappears and a single new substance appears in its place. An example is the historic experiment performed by the German chemist, F. Wöhler, in 1828. Wöhler heated a sample of ammonium cyanate (a compound containing carbon, oxygen, nitrogen, and hydrogen), and thereby converted it into urea, which is an entirely different compound but has exactly the same composition as the starting material. As is well known, urea is also a product of human metabolism. The experiment was the first preparation of an organic compound from an inorganic material and marked the beginning of a new era in the study of organic chemistry. The experiment suggests something interesting for further consideration, namely, that the characteristic properties of a compound depend on more than only the nature and amounts of the elements that compose it. Indeed,

Fig. 3-2
More elaborate apparatus for carrying out a chemical reaction.

we shall see in Ch. 14 that the properties of a compound vary with the three-dimensional pattern according to which the elements are held in combination.

In some cases there is a tendency for reaction to occur, but the reaction does not start unless a stimulus is supplied. For example, zinc powder and sulfur do not undergo a reaction when mixed in the dry state. However, when a lighted match is touched to the mixture, a rapid combination reaction ensues and the compound, zinc sulfide, is formed.

Agents other than heat can also stimulate chemical reactions. For example, a bright light of the appropriate wavelength will trigger the formation of hydrogen chloride gas from a mixture of hydrogen and chlorine. An electrical spark will trigger the reaction of gasoline vapor with air in the combustion chamber of a gasoline engine. In these examples, the heat energy finally released by the chemical reaction far exceeds the energy supplied by the original stimulus.

There are other reactions in which energy must be supplied continuously in order to keep the reaction going. If the supply of energy is suddenly cut off, the reaction stops. For example, the passage of an electric current through water results in its decomposition to the constituent elements, hydrogen and oxygen, but only so long as the electrical current is flowing.

Why these various agents are effective, and why some reactions occur spontaneously while others do not, can be explained theoretically, but not until we have an understanding of the submicroscopic models for elements and compounds. Meanwhile, our discussion continues its rough parallel of the history of chemistry. Having grasped the concept of material substances as composed ultimately of simple elements that combine in certain regular ways under stated conditions, we will find, in the next chapter, that some of the most important laws of chemical combination involve the weights of the combining substances. Since weight is a quality that can be operationally defined, we find ourselves still in the macroscopic world of the chemical laboratory. In the latter part of Ch. 4, we move into the world beyond the reach of direct perception—the submicroscopic world set forth in the atomic theory of John Dalton—to explain the chemical behavior of the macroscopic world.

Suggestions for further reading

Boyle, R., *The Skeptical Chymist* (New York: E. P. Dutton & Co., Inc., 1911).

Flood, H. W., and Lee, B. S., "Fluidization," *Scientific American*, July 1968, p. 94.

Hall, M. B., "Robert Boyle," *Scientific American*, August 1967, p. 96.

King, E. L., *How Chemical Reactions Occur* (New York: W. A. Benjamin Co., Inc., 1963).

More, L. T., *The Life and Works of the Hon. Robert Boyle* (London: Oxford University Press, 1944).

Tilden, W., *Famous Chemists* (New York: E. P. Dutton & Co., Inc., 1921), Chs. 1, 7.

Weeks, M. E., and Leicester, H. M., *The Discovery of the Elements* (7th ed.; Easton, Pa.: Chemical Education Publishing Co., 1968).

Questions

1. (*a*) What is the name for a homogeneous mixture?
(*b*) What two kinds of pure substances are there?
(*c*) How is a compound different from a solution?
(*d*) What are the macroscopic characteristics of chemical change?
(*e*) What is the essential difference between a chemical change and a physical change?

2. (*a*) Describe, in general terms, the following kinds of chemical reaction: decomposition; rearrangement; synthesis; electrolysis.
(*b*) Give an example of each.

3. Decide which of the following changes are physical changes, and which are chemical changes. Explain your answers.
(*a*) The drying of wall paint.
(*b*) The formation of the beverage tea, from hot water and tea leaves.
(*c*) The bleaching of cotton
(*d*) The clotting of blood.

4. A homogeneous solid A melts sharply at 81°C to form a homogeneous liquid B. As B is heated further, decomposition takes place: a gas is evolved and a new solid C appears. Using these facts, decide whether A is an element, a compound, or a solid solution. Then decide whether C is necessarily an element. Explain your answers.

5. A heterogeneous solid A, on being heated to 200°C, changes into a homogeneous liquid B. If the temperature is maintained at 200°C, another change takes place slowly and a *new* solid, C, separates gradually from the liquid phase. When C is isolated and tested, it proves to be a pure substance. Does the change from A to B necessarily involve a chemical reaction? Does the formation of C from B have the characteristic attributes of a chemical reaction? Is C necessarily a compound? Explain your answers.

6. A homogeneous solid A is heated slowly. When the temperature reaches 220°C, decomposition takes place: a gas is evolved, and a new solid, B (which is visibly different from A), appears. B can exist at 220°C indefinitely without decomposition. However, as the slow heating is continued, B begins to decompose noticeably as the temperature approaches 300°C, and when the temperature reaches 300°C, B decomposes completely to form a gaseous product and a silvery solid, C. The new solid, C, can be heated further, even above its melting point of 961°C, without further decomposition. Are these facts sufficient to prove that A and B are compounds and that C is an element? Explain your answers.

SIENA COLLEGE LIBRARY
LOUDONVILLE, N. Y. 12211

The best and safest method of philosophising we should say seems to be, first to inquire diligently into the properties of things, and of establishing these properties by experiment, and then to proceed more slowly to hypotheses for the explanation of them.

ISAAC NEWTON (1642–1727)

Four

THE LAWS OF CHEMICAL CHANGE

IN CONSIDERING the nature of chemical change, the first question we must answer is whether the mass of the products of a chemical reaction is equal to that of the reactants, or whether the mass of the reaction mixture changes as the reaction proceeds. Experiments that hinted at answers to this question were reported as early as 1630 by Jean Rey, a French physician. His findings were later verified and extended by Robert Boyle in 1673. Let us describe some of Boyle's results.

It is well known that iron rusts and copper tarnishes in air at room temperature. These reactions occur much faster at higher temperatures, and Boyle found that his sample of copper changed into a new black compound in a couple of hours when he heated it in a furnace. When he compared the weight of the black compound with that of the original copper, he found that the weight had increased by 6% to 10%. When he heated iron in the same way, the weight increased by 25%; when he heated tin, the weight increased by 12%; and even the "noble" metal, silver, tarnished on heating, and its weight increased by 1%.

The law of conservation of mass

Reading about these results, one might readily conclude that chemical changes are accompanied by an increase in the mass of the material involved. Unfortunately, however, there was a flaw in the design of these experiments: the metals were heated in open vessels, in contact with the air, and it was impossible to tell whether or not the air had participated in the reactions. This aspect of the problem was not understood at the time.

That it was the crux of the matter was recognized at last by Antoine Lavoisier in 1774. Lavoisier sealed a definite amount of air and metal in a glass vessel and weighed the vessel and its contents. He then applied heat until the reaction of the metal seemed to be complete and, without opening the vessel, reweighed it. Since the vessel remained sealed throughout the experiment, matter could neither enter nor leave. Any change in weight, therefore, could be rightfully attributed to the chemical reaction. Moreover, since he used a very sensitive balance, even a small weight change could have been detected. The balance which he had constructed especially for this purpose could detect changes in weight as small as 0.0005 g. (This is roughly 1/100 the weight of a drop

Antoine Laurent Lavoisier:
1743–1794
Lavoisier's careful experimental work and brilliant deductions introduced our modern views of chemical change. [Photograph courtesy Brown Brothers.]

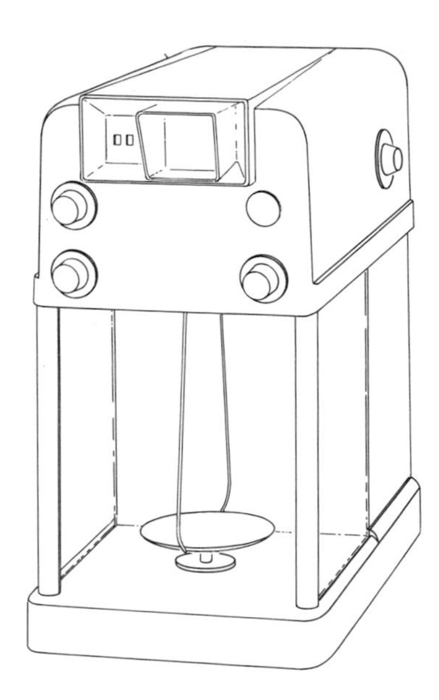

Fig. 4-1
A modern analytical balance.

of water. A modern analytical balance, such as the one shown in Fig. 4-1, will easily weigh to the nearest 0.0001 g.)

In his first experiment, Lavoisier heated a sample of tin above its melting point. The tin tarnished and a black solid product was obtained, readily at first, then more slowly, and after a while the reaction stopped. After the sealed vessel had cooled, it was reweighed. *The weight was exactly the same as before the heating*—definite evidence that no fundamental change in weight had occurred even though there was visible evidence that a chemical reaction had taken place. Lavoisier then opened the retort (see Fig. 4-2) and verified Boyle's result that the black reaction product weighed more than the tin from which it had been obtained. But at the same time he found that the weight of the air inside the closed vessel had decreased by an equal amount. In modern terminology, what happened was that the tin had reacted with the oxygen in the air to form a new substance, tin oxide.

Lavoisier's finding that the weight of the reaction products is equal to that of the reactants which have disappeared and that the total weight remains constant has been confirmed many times for all sorts of chemical reactions, and no significant exceptions have ever been noted. Thus, we may formulate a general rule that summarizes all observations of this kind. We shall call this rule the LAW OF CONSERVATION OF MASS:

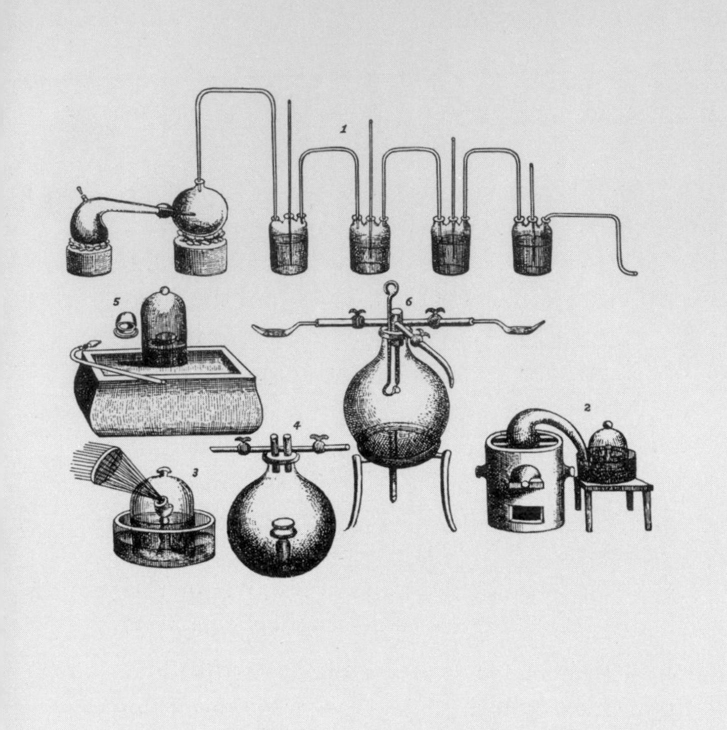

Fig. 4-2
Some of Lavoisier's chemical apparatus. (1) Apparatus for the production of gaseous substances. The desired gas is produced by heating a suitable substance in the *retort* on the left, and then purified by bubbling through a series of gas-washing bottles. (2) Apparatus used in the oxidation of mercury. (3) and (4) Apparatus for the combustion of phosphorus. Note the lens or *burning glass* which concentrates solar heat on the sample. (5) The same apparatus, with pneumatic trough of marble. (6) Apparatus for the synthesis of water using electric spark.

Matter is neither destroyed nor created when a chemical reaction takes place: the mass of the reaction products is precisely equal to the mass of the reactants from which the products are obtained.

This law is fundamental. All of analytical chemistry is based upon it. Since chemical analysis is being done continually by many people, in laboratories all over the world, one can truthfully say that the law of conservation of mass is still being tested every day.†

The elementary analysis of compounds

In Ch. 3 we stated that when two pure substances combine to form a given compound, they do so in a definite proportion by weight. We shall now discuss the experimental evidence in support of that state-

† The law of conservation of mass applies only to ordinary chemical reactions, and not to the reactions of atomic nuclei. This limitation of the law will be discussed in Ch. 18.

ment. Our remarks will also serve as an introduction to the methods of chemical analysis.

The easiest and most direct method of determining the composition of a compound is to find the weights of the elements that must react in order to form the compound. A good example is found in the work of the Swedish chemist, J. J. Berzelius (1779–1848), on the composition of lead sulfide. Ten grams of the element, lead, were heated with three grams of the element, sulfur, until reaction between the two was complete. The reaction product was lead sulfide, a black crystalline solid that was easily distinguished from lead, a soft silvery metal that turns grey on exposure to air, and from sulfur, a yellow solid. Since the two elements and their product are so different in appearance, it was easy for him to discern that the lead had reacted completely, while some of the sulfur was left unreacted. After removing the unreacted sulfur by solvent extraction with carbon disulfide, a liquid in which lead sulfide is very insoluble, he dried and weighed the pure reaction product and found that 11.56 g had been formed. From these data he deduced the composition of lead sulfide as follows:

Wt. of lead sulfide:	11.56 g	
Wt. of lead:	10.00 g	(equal to the weight of lead taken originally, since all of it has reacted)
Wt. of sulfur:	1.56 g	(by difference)
Wt. % of lead:	$\frac{10.00}{11.56} \times 100 = 86.5$	
Wt. % of sulfur:	$\frac{1.56}{11.56} \times 100 = 13.5$	

On repeating the analysis, Berzelius found that the composition of lead sulfide (that is, the weight per cent of lead and of sulfur) was always the same, regardless of how much lead and how much sulfur was allowed to react. If more of either lead or sulfur was used than was necessary for the formation of lead sulfide, then the extra amount simply remained, unreacted.

Another method of finding the composition of a compound is to decompose it into its constituent elements, and to measure the amount of each. For example, the compound mercuric oxide, a red solid, is decomposed by heating into the constituent elements, mercury (a liquid) and oxygen (a gas). Upon complete decomposition, 45.0 g of mercuric oxide yields 41.5 g of mercury. From these data the composition of mercuric oxide is computed as follows:

Wt. of mercuric oxide:	45.0 g	
Wt. of mercury combined therein:	41.5 g	(weighed after decomposition)
Wt. of oxygen:	3.5 g	(by difference)
Wt. % of mercury:	$\frac{41.5}{45.0} \times 100 = 92.2$	
Wt. % of oxygen:	$\frac{3.5}{45.0} \times 100 = 7.8$	

In a third method of chemical analysis, the compound of unknown composition is allowed to react with another element to form a new compound whose composition has already been determined. For example, 10.00 g of sodium chloride in solution will react with silver nitrate in solution to form 24.52 g of a new, insoluble compound, silver chloride, whose composition is known as 75.26 wt. % silver and 24.74 wt. % chlorine. In this reaction, all of the chlorine that originally

Jöns Jakob Berzelius: 1779–1848
Berzelius' painstaking analytical work did much to establish the validity of the atomic theory of matter. [Photograph courtesy Brown Brothers.]

combined with the sodium in sodium chloride combines with silver to form silver chloride. The composition of sodium chloride may now be computed by the following method:

Wt. of silver chloride obtained:	24.52 g	
Wt. of chlorine in 24.52 g of silver chloride:	$24.52 \times \frac{24.74}{100} = 6.07$ g	
Wt. of chlorine in 10.00 g of sodium chloride:	6.07 g	(same as in silver chloride)
Wt. of sodium in 10.00 g of sodium chloride:	3.93 g	(by difference)
Wt. % of sodium in sodium chloride:	39.3	
Wt. % of chlorine in sodium chloride:	60.7	

The law of definite composition

Now that we know how the composition of a substance is determined, we are ready to learn a most important generalization:

Whenever two or more elements combine to form a given compound, they do so in a definite proportion by weight.

This rule is known as the LAW OF DEFINITE COMPOSITION. Since we identify the given compound by its physical properties, another way of stating this law is:

Whenever two samples of a pure substance are identical in their physical properties, they are also identical in their chemical composition.

The law of definite composition applies regardless of the source of the samples or the method of analysis that is used. Thus, the chemical composition of a given pure substance is just as definite and characteristic a property of the substance as any of the physical properties, and it may be used in the same way to identify the substance.

It is a fact, however, that two different substances may fortuitously have the same composition, just as two different substances may happen to have the same density or melting point. The identification of a substance should, therefore, never be based on its composition alone, just as it should never be based on any single physical property. However, measurement of two or three physical properties in addition to the composition should be sufficient for identification.

Dalton's atomic theory

We are now ready to depart from the macroscopic examination of matter and seek an explanation in terms of a submicroscopic model. The two *macroscopic* laws, conservation of mass and definite composition, are all explained elegantly by Dalton's atomic theory of matter (1808).† In its time Dalton's theory was rather a bold guess because the experimental basis for those laws was still fairly unsatisfactory. However, Dalton accepted the laws as correct because his own experiments seemed to confirm them and because they accorded with his intuitive faith in the underlying simplicity of nature.

According to Dalton's theory, matter that appears homogeneous and continuous on the macroscopic scale is nevertheless *dis*continuous on the submicroscopic scale, being made up of large numbers of particles. This becomes clear when we consider Dalton's postulates, phrased in modern language.

1. THE PARTICLE NATURE OF THE ELEMENTS:

a. A sample of an element consists of many tiny particles called *atoms*.
b. Atoms cannot be divided nor destroyed nor created.
c. Atoms of the same element are alike in all respects.
d. Atoms of different elements are characteristically different, particularly in their weights.

2. THE NATURE OF COMPOUNDS:

a. A sample of a compound consists of many tiny particles called *molecules*.††
b. All molecules of the same compound are alike in all respects.
c. When a compound is decomposed into the constituent ele-

† The word "atom" is derived from a Greek root meaning "indivisible." Originally, an atom was thought to be a particle of matter so small as to be incapable of further division.

†† Derived from the Latin word *molecula*, meaning "small amount of matter."

ments, the molecules of the compound break up to form the elementary atoms.

d. When a compound is formed from the elements, the atoms of the elements combine *in whole number amounts* to form the molecules of the compound. This is because the atoms are indivisible.

e. Atoms of different elements can combine in more than one ratio to form different molecules and hence, different compounds.

Several of these ideas will have to be modified in later chapters. But for the present, this simple theory is useful, since it explains the two laws that were presented above. First, the theory explains the law of conservation of mass because atoms cannot be created or destroyed. When a

John Dalton: 1766–1844
In 1808, Dalton published his celebrated ***New System of Chemical Philosophy,*** in which he developed his conception of atoms as the fundamental building blocks of all matter. [Photograph courtesy Brown Brothers.]

reaction takes place, the atoms change their state of combination with other atoms; for example, atoms of hydrogen may combine with atoms of chlorine to form molecules of hydrogen chloride. But the total number of hydrogen atoms and of chlorine atoms and, hence, the weight of the reaction mixture, remain the same.

The theory also explains the law of definite composition. For example, in each molecule of hydrogen chloride there is some definite number of hydrogen atoms combined with some definite number of chlorine atoms. These numbers are the same in every molecule, and they are always whole numbers. Since each molecule has exactly the same composition, any collection of them will also have the same composition. Thus, the theory explains the observation that all samples of pure hydrogen chloride have exactly the same composition.

CHEMICAL SHORTHAND

Dalton introduced the practice of using symbols to denote the atoms of the different elements. This practice leads to a convenient and concise way of representing the composition of molecules.

The symbols that are in use today were originated by Berzelius. For most elements, the first letter of the chemical name of the element is employed to denote the atoms of that element. For example, the letter H is used for the atoms of hydrogen, O for the atoms of oxygen, N for the atoms of nitrogen, C for the atoms of carbon, and S for the atoms of sulfur. Since the notation is used throughout the world, the symbol denoting an element is sometimes the first letter of the name of the element in some language other than English. For example, K, the symbol for the element potassium, derives from the word *kalium,* which is of mixed German–Latin origin.

When a single letter might be confused with the letter symbol used for some other element, two letters are used. For example, atoms of the element helium are denoted by the symbol He to avoid confusion with the symbol H for atoms of hydrogen. Na, from the Latin word *natrium,* is used for the atoms of sodium; Cu (*cuprum*) is the symbol for copper, Cl is the symbol for chlorine, and Fe (*ferrum*) is the symbol for iron.

The chemical formula for the molecules of a compound is obtained as follows: the symbols are given for the elements of which the compound is composed, with a subscript following each symbol to denote the number of atoms (if more than one) of that element in one molecule of the compound. For example, the formula CO denotes a molecule consisting of one carbon atom and one oxygen atom; CO_2 denotes a molecule consisting of one carbon atom and two oxygen atoms; and C_7H_{16} denotes a molecule consisting of seven carbon atoms and sixteen hydrogen atoms.

The same kind of symbols can be used also to depict the overall course of a chemical reaction. This is done by writing down the formulas of the reactants and products, with an arrow to indicate that a reaction has taken place. The result is called a *chemical equation.* As an example of a very simple chemical equation, consider the reaction described on p. 45, in which lead sulfide is formed by heating together lead and sulfur. Since the symbol for lead is Pb (plumbum) and that for sulfur is S, we may write:

$$\underset{\text{(Reactants)}}{Pb + S} \longrightarrow \underset{\text{(Product)}}{PbS}$$

This chemical equation is read as follows:

Lead plus sulfur *gives* lead sulfide.

In writing a chemical equation, it is necessary to comply with Dalton's postulate that atoms can neither be divided nor created nor destroyed. That is, for each species of atoms, the number of atoms in the reactant molecules must be equal to that in the product molecules. The process of ensuring that the numbers of atoms are equal is called *balancing the equation.* Procedures for doing so will be discussed in Ch. 7.

The law of multiple proportions

The combination of two elements will not always produce a single compound. This is one of those ways that nature complicates chemistry, but makes it more interesting, too. When more than one compound is formed, the relative weights of the two elements that combine to form the different compounds always conform to a rather simple rule. This rule is called the LAW OF MULTIPLE PROPORTIONS:

When two elements, A and B, combine to form two different compounds, the weights of B in the two compounds, combined with a fixed weight of A, are in the ratio of small whole numbers.

An example is furnished by the oxides of carbon. Two oxides are commonly known: the highly poisonous gas, carbon monoxide, and the gas, carbon dioxide, which is usually produced when carbon or one of its compounds burns in oxygen or in air. Pure carbon monoxide may be prepared by passing carbon dioxide gas over red-hot charcoal, which is a slightly impure form of carbon. The two gases are readily distinguished by their different chemical properties. For example, carbon dioxide is absorbed completely by the solid substance, potassium hydroxide, whereas carbon monoxide is not.

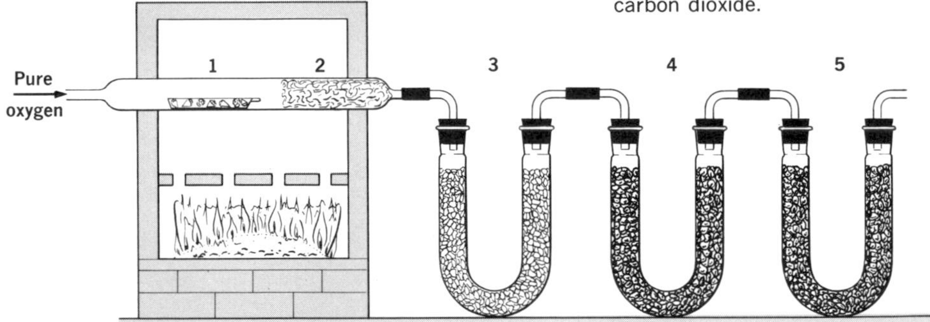

Fig. 4-3
Apparatus for the elementary analysis of carbon dioxide. (1) Platinum boat containing the diamonds. (2) Solid copper oxide, to convert possible traces of carbon monoxide to carbon dioxide. (3) Solid "drying agent" to remove traces of water. (4) Potassium hydroxide, to absorb carbon dioxide. (5) Potassium hydroxide, to guard against access of atmospheric carbon dioxide.

The accurate analysis of the two gases proved to be a difficult problem in the early days of quantitative chemistry. It was finally accomplished in a painstaking series of experiments carried out between 1841 and 1849 by the French chemist, J. B. Dumas, and his Belgian coworker, J. S. Stas. To obtain an accurate analysis of carbon dioxide, a stream of dry oxygen gas was passed over diamonds that were heated to a red heat. Diamonds were used because they were the purest source of carbon then known. At the high temperatures, the diamonds reacted slowly with the oxygen to form carbon dioxide gas, which was swept away by the stream of oxygen into another part of the apparatus where it was absorbed by solid potassium hydroxide. The weight of carbon dioxide was determined by weighing the absorbent before and after the experiment, and the weight of carbon in that much carbon dioxide was, of course, equal to the weight of diamonds that had reacted. Thus, the composition of carbon dioxide could be computed.

A sketch of the type of apparatus used is shown in Fig. 4–3. The apparatus proved so satisfactory that, with only minor modification, the same type of apparatus is still used to determine the per cent of carbon in carbon compounds.

To analyze carbon monoxide, a sample of the pure gas was allowed to react with a known weight of oxygen; the product, carbon dioxide, was absorbed by potassium hydroxide and weighed. Since the composition of carbon dioxide was already known, the composition of carbon monoxide could be computed as explained on page 45.

The results reported by Dumas and Stas were as follows:

Carbon monoxide:	1.000 g of carbon is combined with 1.336 g of oxygen
Carbon dioxide:	1.000 g of carbon is combined with 2.667 g of oxygen

Thus the amounts of oxygen in combination with a fixed weight (1.000 g) of carbon are in the ratio 2.667:1.336, or 1.995:1. Allowing for possible experimental error, this ratio could be exactly 2:1, a ratio of small whole numbers.

There are many other examples of similar simple relationships when two elements combine to form more than one compound. A few of them are summarized in the following table.

EXAMPLES OF MULTIPLE PROPORTIONS

Elements	Compound	Composition	Ratio	Nearest ratio of small whole numbers
Lead and oxygen	Brown oxide: Yellow oxide:	100 g : 15.6 g 100 g : 7.8 g	2.00:1	2:1
Sulfur and oxygen	Sulfur trioxide: Sulfur dioxide:	100 g : 146.427 g 100 g : 97.83 g	1.497:1	3:2
Copper and oxygen	Black oxide: Red oxide:	100 g : 25 g 100 g : 12.5 g	2.00:1	2:1
Iron and oxygen	Ferric oxide: Ferrous oxide:	100 g : 44.25 g 100 g : 29.6 g	1.495:1	3:2

The data in this table are based on the work of J. J. Berzelius. Within limits consistent with experimental error, the weights of the second element in combination with 100 g of the first are always in the ratio of small whole numbers.† Berzelius was a magnificent experimentalist and was able to show that certain apparent exceptions to the rule, reported by others, came from the use of impure substances or experimental error. As a result, by the middle of the nineteenth century the law of multiple proportions was generally accepted.

Only later, as chemistry developed and chemists learned to prepare compounds of greater molecular complexity, did the law have to be modified. The word "small" had to be removed from the phrase "small whole numbers" in order for the law to apply also to compounds with large molecules. For example, ordinary gasoline consists mostly of various compounds of the elements carbon and hydrogen, the so-called hydrocarbons. In two of these compounds, namely in heptane and octane, the amounts of hydrogen in combination with 100 g of carbon are in the ratio 64 : 63. This is still the ratio of two integers, but the integers are no longer small numbers.

Finally, it must be mentioned that there is a class of substances, which includes the zeolites and the gels, that are characterized by a

† Experimental error results in part from the limitations in the sensitivity of measuring instruments, and in part from poor experimental technique.

highly porous structure and large surface area. Materials such as Portland cement concrete, gelatin, and water softeners are in this group. These substances can adsorb large quantities of matter on their surfaces, and their composition can thus be varied continuously.

The law of multiple proportions is easily understood in terms of Dalton's atomic theory. For example, let us consider the two oxides of carbon. The simplest conceivable compound of these elements has the formula CO; that is, in each molecule there is one atom of carbon and one atom of oxygen. Another logical possibility is a compound molecule with the formula CO_2. If we consider samples of such oxides containing an equal number of carbon atoms, then it is obvious that the sample of CO_2 will contain precisely twice as much oxygen as that of CO.

Although the suggested formulas, CO_2 and CO, agree with the 2 : 1 ratio of the data on p. 53, there are other pairs of formulas that agree equally well. For example, a 2 : 1 ratio is consistent also with CO and C_2O, or with CO_4 and CO_2, or with any other pair of molecular formulas in which one formula shows twice as many oxygen atoms per carbon atom as does the other. This multiplicity of possible molecular formulas suggests one of the shortcomings of Dalton's theory: the theory cannot predict molecular formulas uniquely; some additional information or postulate is needed. In trying to solve this problem, Dalton adopted the viewpoint that nature is intrinsically simple, and he guessed, therefore, that common, stable compounds are formed from their constituent elements by simple one-to-one combination of atoms. Thus, in the case of carbon monoxide, he assumed that the molecular formula is CO, which happened to be correct. In the case of water, he assumed that the molecular formula is HO, which (as will be shown in Ch. 6) happened to be wrong.

Facts, laws, and theories

The products of scientific endeavor are experimental facts, natural laws, and theories. These terms are used a great deal in scientific discussions, and it is surprising, therefore, and unfortunate that there is no universal agreement on just what they mean. The definitions employed by the authors and by many other scientists can be stated as follows.

If a process or object can be perceived directly by the human senses, a statement about it that has been verified by these senses is called a *fact*. Consequently, verified observations of physical or chemical properties, or of macroscopic processes involving matter, are called facts.

A *natural law* is a generalization based upon a large body of facts related to one phenomenon or process. It may take the form of a concise statement, or it may be a mathematical equation. In either case, it summarizes all previous observations and predicts that, under the same circumstances, the same observations will be made in the future. For example, the law of definite composition is the simple statement that any two samples of the same pure substance always contain the same elements in the same proportion by weight. This statement is consistent with all past experience about the composition of substances, and it predicts that the same sort of behavior will be found in all future investigations. However, since scientists are frequently redetermining the composition of known substances and are preparing new substances, some of which may be radically different from those known today, it is not impossible that some day something will turn up that is not consistent with this law. If this happens, the law will have to be modified, or restated, or perhaps even abandoned. Thus, the very quality of prediction which makes a law useful to the scientist also makes it vulnerable, for a single inconsistent observation may necessitate revision.

Laws are formulated in the belief that there is a fundamental orderliness to nature so that, under the same set of circumstances, the same results will be observed again and again. Even when an observation is inconsistent with an existing law, the scientist is not likely to lose his faith in the "lawfulness" of nature. He will search for a new, more general law that is consistent also with the inconsistent observation, or he will restate the law so that it no longer applies to this observation. For example, there was a time when the law of conservation of mass was thought to apply to all the changes that matter undergoes. Then it was found that this law fails when it is applied to nuclear reactions. As a result, a more general law of conservation of mass-energy has replaced it; the new law applies to both chemical and nuclear reactions. And the law of conservation of mass has been restated so that it applies in chemical reactions because the change in mass, while presumably real, is too small to be detected.

As has been pointed out, scientists are not satisfied only to observe facts and formulate laws, but they try also to explain the facts and the laws. Because they are unable to perceive the innermost workings of matter directly, they *invent* possible explanations involving submicroscopic particles and events. These explanations by means of submicroscopic models are called *theories*.

There are certain ideals that scientists have in mind when they invent theories. In the first place, the ideal theory must be able to explain all known relevant facts: the entire macroscopic world. In the second place, it must be as simple as possible. The strict discipline imposed by these two ideals prevents scientific theory from becoming wild speculation. To be acceptable, a theory must be able to explain a substantial body of facts; it will not do to invent a special theory to explain each individual fact. Furthermore, a theory must be clear and simple, so that when it is applied to a new situation it will readily lead to an unambiguous prediction; it is not satisfactory to populate the submicroscopic world with unpredictable figures like demons or elves whose behavior in a new situation is anybody's guess.

Unfortunately, the two ideals, economy of theory and simplicity of statement, are not entirely compatible. For example, a simple pictorial model may succeed in explaining an entire field of knowledge but be unsuccessful when used in another field. If the more general theory which explains both fields is much more complicated and harder to pictorialize, it is advantageous to continue to use the simple theory whenever it applies. Such a situation actually exists in chemistry. The most general theory is highly abstract, mathematical, and difficult to pictorialize. But fortunately, most of chemistry can be explained by means of simple extensions of Dalton's theory in which each molecule may be visualized as a concrete entity, with a definite geometrical structure and a fixed position for each atom. Chemists actually build large-scale models of these submicroscopic molecules to help them visualize chemical reactions. See, for example, the figures on pp. 256–260.

A theory in chemistry can never become a law because one can never hope to observe directly the particles and processes which the theory postulates. For example, even though there is now a tremendous amount of evidence in support of the atomic theory of matter, the theory can never become a law simply because we can never hope to perceive atoms directly through the human senses.†

Bertrand Russell has given an illuminating account of the stages in the gradual maturing of a scientific theory: "Every physical theory which survives goes through three stages. In the first stage, it is a matter of controversy among specialists; in the second stage, the specialists are agreed that it is the theory which best fits the available evidence, though it may well hereafter be found incompatible with new evidence; in the third stage, it is thought very unlikely that any new evidence will do more than somewhat modify it."††

Suggestions for further reading

Brown, J. F., Jr., "Inclusion Compounds," *Scientific American,* July 1962, p. 82.

Dinga, G. P., "The Elements and the Derivation of their Names and Symbols," *Chemistry, 41,* February 1968, p. 20.

Duveen, D. I., "Lavoisier," *Scientific American,* May 1956, p. 84.

Lowry, T. M., *Historical Introduction to Chemistry* (London: Macmillan & Co. Ltd., 1936).

McKie, D., *Lavoisier* (New York: Henry Schuman, Inc., 1952).

Neville-Polley, L. J., *John Dalton* (New York: The Macmillan Company, 1920).

Szabadvary, F., "A Visit with Antoine Lavoisier," *Chemistry, 42,* April 1969, p. 14.

Tilden, W., *Famous Chemists* (New York: E. P. Dutton & Co., Inc., 1921), Chs. 6, 8, 11.

Questions

1. Lithium perchlorate is a compound of the elements lithium, chlorine and oxygen. When lithium perchlorate is heated above 430°C, decomposition occurs: oxygen gas is evolved, and solid lithium chloride (a compound of lithium and

† Some scientists, however, have so much confidence in a well-established theory that they do call it a law.

†† Bertrand Russell, *Human Knowledge* (New York: Simon and Schuster, Inc., 1948), p. 198.

chlorine) is produced. 10.00 g of pure lithium perchlorate on decomposition yields 3.98 g of lithium chloride.
(*a*) Calculate the weight of oxygen gas that is evolved.
(*b*) Calculate the weight percent of oxygen in lithium perchlorate.

2. The element chlorine, when irradiated with ultraviolet light, reacts with the element hydrogen in the gas phase to produce the compound, hydrogen chloride. On reacting completely, 1.240 g of chlorine gives rise to 1.275 g of hydrogen chloride.
(*a*) Find the weight of hydrogen that has reacted.
(*b*) Find the weight percentages of the elements, hydrogen and chlorine, in the compound, hydrogen chloride.

3. 1.00 g of magnesium is mixed with 5.00 g of chlorine. A reaction takes place which results in the formation of the compound, magnesium chloride. The magnesium is used up completely, but it is found that 2.05 g of chlorine is left unreacted. What is the weight percent of magnesium in magnesium chloride?

4. Express in chemical shorthand:
(*a*) A molecule consisting of six carbon atoms and six hydrogen atoms.
(*b*) A molecule consisting of two carbon atoms and four hydrogen atoms.
Are the molecules represented in (*a*) and (*b*) consistent with the law of multiple proportions?

5. Give a submicroscopic description of a decomposition reaction, based on Dalton's atomic theory.

6. Consider the examples of multiple proportions given in the Table on p. 54.
(*a*) For the compounds of lead (Pb) and oxygen, if the molecular formula of the yellow oxide is PbO, what is a possible formula for the brown oxide?
(*b*) For the compounds of sulfur and oxygen, if the molecular formula for sulfur dioxide is SO_2, what is a possible formula for sulfur trioxide?
(*c*) For the compounds of iron (Fe) and oxygen, if the molecular formula for ferrous oxide is FeO, what is a possible formula for ferric oxide?

7*. Express the following by means of a chemical equation:
(*a*) An atom of barium (Ba) reacts with an atom of sulfur (S) to form a molecule consisting of one barium atom and one sulfur atom.
(*b*) A molecule consisting of one carbon atom, two oxygen atoms and two hydrogen atoms decomposes into a molecule consisting of one carbon atom and one oxygen atom, *plus* a molecule consisting of two hydrogen atoms and one oxygen atom.

*Questions with an asterisk (*) are more difficult or require supplementary reading.

How awkward is the human mind in divining the nature of things, when forsaken by the analogy of what we see and touch directly.

LUDWIG BOLTZMANN (1844–1906)

Five

ENERGY AND THE KINETIC MOLECULAR THEORY OF MATTER

WE HAVE seen that Dalton's theory, according to which matter consists of tiny particles or molecules, can explain many quantitative facts concerning chemical reactions. In this chapter, we wish to consider another and independent way of arriving at a particle theory of matter. Our discussion will deal with the *physical* changes that matter undergoes as it is heated or cooled, and particularly with the amounts of energy that are absorbed or evolved when matter changes its temperature or physical state.

Heat and work

There are two methods for raising the temperature of a substance: heating it or doing work on it. The work may involve the vigorous stirring of a fluid, or the rubbing of a solid, or the compressing of a gas, for example. Conversely, if we wish to lower the temperature of a substance, we can allow heat to flow out of it into the surroundings, or we can arrange things so that the substance performs work. For example, an expanding gas might do work by driving the piston of an engine.

Physicists have developed methods for measuring the amount of heat flowing into or out of a substance, as well as the amount of work done on or by a substance. The heat is commonly expressed in terms of a unit called the *calorie,*† which is defined as follows:

One calorie is the amount of heat required to raise the temperature of one gram of water one degree centigrade, from 14.5°C to 15.5°C.

Although this precise definition specifies that the temperature must be raised from 14.5 to 15.5°C, one calorie is also equal, in good approximation, to the amount of heat required to raise the temperature of one gram of liquid water one degree at any temperature between 0°C and 100°C. Therefore, if we wish to raise the temperature of 1000 g of water 5°C, 5000 calories must be supplied. Or if we wish to lower the temperature of 200 g of water from 25°C to 22°C, 600 calories of heat must be removed.

There are several units in use for measuring the amount of work done, but it would take us too far afield to specify their definitions. One such unit, which is well-known to users of electricity, is the kilowatt-hour. Since it is possible to raise the temperature of a substance by doing work on it rather than by causing heat to flow into it, it seems reasonable to inquire whether there is a relationship between the amounts of heat and work required to bring about a given temperature change. That such a relationship indeed exists was shown more than a hundred years ago by James Prescott Joule, an English physicist (1818–1889). Joule found that the addition of one calorie of heat is always equivalent in its effect to the performance of a certain definite quantity of work. (For example, one calorie is equivalent to 1.16×10^{-6} kilowatt-hour, if that is the unit chosen to express the work.) This relationship is true regardless of the nature of the substance, whether it be solid, liquid, or gaseous, and regardless of the detailed manner in which the heat is supplied or the work is done. Indeed, the relationship is so accurate that quantities of heat and work can both be expressed in the same units, calories.

† From the Latin "calor," which means heat.

Energy

Because of the equivalence of the effects of heat and work, scientists have invented a more general concept that encompasses both. This is the *energy* of a substance. Consider that the temperature of a substance is raised either by adding one calorie of heat or by doing an equivalent amount, one calorie, of work. The final state of the substance will be the same. We can therefore give a completely adequate description of the result by saying merely that the energy of the substance has increased by one calorie, without specifying whether the energy was added in the form of heat or work.

It may be helpful to consider the following analogy. Suppose Mr. Jones wants to deposit $100 in his bank account. He might carry the amount to the bank in the form of a hundred dollar bill, or he might tote a sackful of pennies to the bank. The manner in which the deposit is made is evidently quite different in the two cases. Yet the final result of a deposit is exactly the same, since the bank account has no memory of the form in which the money was added to it.

Similarly, we can say that a substance in a given state has a certain, definite amount of energy "deposited" in it. Yet the substance has absolutely no memory of the manner in which the energy was deposited. When one calorie of energy enters the substance because heat is absorbed or work is done, we say that the energy of the substance has increased by one calorie. When one calorie of energy leaves the substance because heat is given off or work is done by the substance, we say that the energy of the substance has decreased by one calorie. The increase or decrease of energy is always accompanied by a perceptible change in the properties of the substance. We have already mentioned that the temperature might change. Some other possibilities, such as the melting of a solid or the freezing of a liquid, will be considered in the next section.

The energy of a substance is increased not only by the absorption of heat or the performance of mechanical work, but also by the absorption of electrical energy, or light, or sound. When all possible forms of energy are considered, it is found that the total number of calories absorbed by a substance is precisely equal to the total number of calories given up by the surroundings. Conversely, the total number of calories given off by a substance is exactly equal to the total number of calories received by the surroundings. These precise relationships are summarized by the LAW OF CONSERVATION OF ENERGY:

Energy is neither created nor destroyed in any process; the total amount of energy of the substance undergoing the process, plus that of the surroundings, remains constant.†

The behavior of pure substances as energy is added or removed

Energy is required not only to raise the temperature of a substance, but also to bring about a change of state, such as from solid to liquid or from liquid to vapor. It is common knowledge, for example, that heat must be supplied to cause ice to melt at 0°C, or water to boil at 100°C. Thus, ice stored in a thermos flask melts rather slowly, because the energy required to convert it to liquid water can leak through the insulation only slowly. Or, when a kettle of boiling water is taken off the flame, boiling stops at once after the supply of heat has been terminated. These facts show that liquid water at 0°C has more energy than ice at 0°C, and that steam at 100°C and 1 atm has more energy than liquid water at 100°C and 1 atm. The behavior of water in this respect is not at all unusual. *Any* solid at its melting point has less energy than the liquid that is produced; and any liquid at its boiling point has less energy than the vapor that is produced.

The relationship between the energy of the solid, liquid, and gaseous state of a pure substance is shown schematically in Fig. 5-1, which shows the temperature of the substance as energy is added to it. The line AB corresponds to the heating of the solid; here the temperature increases as energy is added. Line BC corresponds to the melting of the solid. Line BC is horizontal because for pure substances melting occurs at constant temperature. Line CD corresponds to the heating of the liquid; here again the temperature increases as energy is added. Line DE corresponds to the boiling of the liquid at atmospheric pressure. The temperature remains constant until all the liquid is converted to

†See, however, discussion of the law of conservation of mass-energy on p. 56.

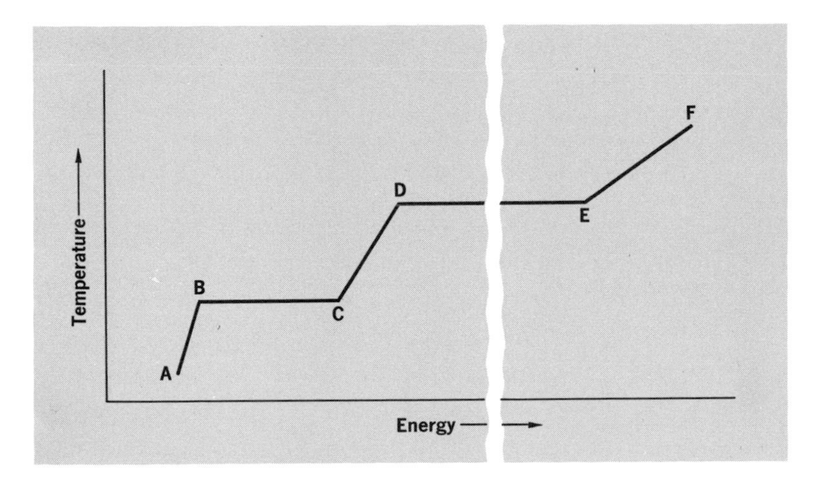

Fig. 5-1
Diagram showing how the temperature varies with the energy for a typical pure substance. Line DE is truncated to about one-tenth of its typical length.

vapor. The amount of energy required for vaporization is usually much greater than that required for melting. Finally, line EF represents the heating of the vapor.

Kinetic and potential energy

We have seen that energy enters or leaves a substance when the substance changes its temperature or its state. So far, we have not tried to visualize what happens to the energy inside the substance. In the remainder of this chapter we will show that a particle theory of matter can give a satisfactory explanation. In such a theory it is assumed that the submicroscopic particles are in motion, and that there may be forces of attraction and repulsion acting between them. The energy of the macroscopic substance is simply the sum total of all the energy that results from these submicroscopic motions and interactions. In order to understand how such a collection of submicroscopic particles can possess energy, we will consider first the energy that *macroscopic* particles possess by virtue of their macroscopic motion or interaction.

KINETIC ENERGY

Moving objects are capable of doing work. They must therefore possess energy. For example, a falling hammer can drive a nail, or a gust of air can drive a windmill. The energy that macroscopic samples of matter possess by virtue of their motion is called *kinetic* energy.

In measuring the kinetic energy of a rigid body, we say arbitrarily that the kinetic energy is zero when the body is at rest. By "at rest" we mean that the body is stationary with respect to a convenient reference point, usually a fixed point on the surface of the earth. The rigid body is in motion when it is moving toward or away from an observer who is stationed at the reference point and is looking in a fixed direction, or when it is spinning, as viewed from the reference point. The kinetic energy can then be defined as the total quantity of energy that appears, either as heat or as work or in any other form, when the moving body is suddenly brought to rest.

Kinetic energy is divided into two kinds, depending on the type of motion: kinetic energy due to forward motion is called *translational* energy; and kinetic energy due to rotation about an axis or axle is called *rotational* energy. The latter includes the kinetic energy that a spinning object possesses because of its spin.

The translational energy is given by the formula

$$\text{Translational energy} = \tfrac{1}{2} \cdot mv^2$$

where m is the mass of the moving body, and v the velocity of its forward motion. Note that the energy increases as the square of the velocity, so that objects moving at high speeds have very high kinetic energies as compared to the same objects moving at low speeds. Translational kinetic energy is inherent also in vibration, or back-and-forth motion.

The rotational energy is proportional to the square of the speed of rotation, which is measured by the number of revolutions per second.

$$\text{Rotational energy} = \tfrac{1}{2} \cdot I \cdot (\text{Speed of rotation})^2$$

The quantity I appearing in this equation is called the *moment of inertia* and is a measure of the inertia of the body toward rotation about the given axis. It is analogous to the mass, m, in the expression for the translational energy, inasmuch as the mass is a measure of the inertia toward straight-line motion.

POTENTIAL ENERGY

If an object is moving in isolation, it will continue to move indefinitely in a straight line and with constant velocity. However, if other material bodies are present that attract or repel it, the direction and/or velocity of the motion will change. For example, in the solar system, the planets travel in curved paths about the sun rather than in straight-line paths because of the gravitational attraction between the sun and each planet. Gravitational attraction is a general phenomenon that exists between any two material objects; the force of attraction varies in direct proportion to the mass of each object and in inverse proportion to the square of the distance of separation.

Another interaction is that due to electrical charges.† If the kinds of electrical charge on the two objects are different (one positive and one negative), the objects attract each other; if they are alike (both positive or both negative), the objects repel each other. For fixed amounts of charge, the force of attraction or repulsion decreases as the inverse square of the distance between the charges.

Another familiar example is the attraction between a permanent magnet and iron filings. Here the attraction is a magnetic one, and again the force decreases with increasing distance.

The energy that a collection of objects or particles possesses as a result of attractive or repulsive interactions between the particles is called *potential energy*. Any interaction between two particles, whether it be attractive or repulsive, has the property that it becomes small at large distances and approaches zero at very large distances, since the interaction must necessarily vanish as the particles become isolated. We therefore define potential energy so that it becomes zero as the particles move infinitely far apart.

†See page 123 for further discussion of the concept of electrical charge.

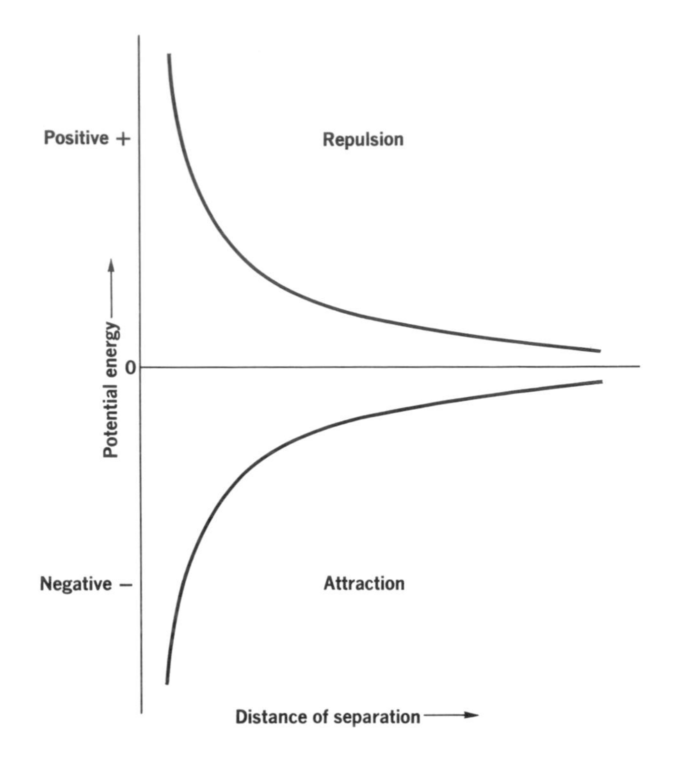

Fig. 5-2
Variation of potential energy with distance of separation.

Figure 5-2 shows qualitatively how the potential energy of two particles varies with the distance of separation in two special cases. The upper curve illustrates the case where two particles *repel each other at all distances.* At very large distances, the potential energy is zero, by definition. Because of the mutual repulsion, the particles will not approach each other spontaneously, but work must be done against the repulsive forces to bring the particles closer together. The upper curve in Fig. 5-2 is therefore drawn so that the potential energy *increases* as the particles are brought closer together. The increase in potential energy is precisely equal to the work that must be done.

The lower curve in Fig. 5-2 illustrates the case where the two particles *attract each other at all distances.* At very large distances, the potential energy is again zero, by definition. However, because of the attractive forces, the particles tend to approach each other spontaneously. Consequently, it is possible to bring about the approach of the particles in such a way that useful work is done in the surroundings. Accordingly, the lower curve is drawn so that the potential energy *decreases* as the particles are brought closer together. The decrease in potential energy is precisely equal to the quantity of useful work that can be done. Since the potential energy decreases from an initial value of zero at very large distances, it must assume negative values at smaller distances. This is brought out in the figure, since the lower curve never goes above the *x*-axis.

The preceding examples illustrate the manner in which potential-energy curves are drawn. When the net force acting between the particles is one of attraction, the potential energy decreases as the particles are brought closer together. When the net force is one of repulsion, the potential energy increases as the particles are brought closer together. Oftentimes the nature of the force changes with the distance. For example, Fig. 5-3 shows how the potential energy varies with distance when there is a net force of attraction at long distances and a net force of repulsion at short distances. At long distances, the potential energy decreases consistently (becomes more negative) as the distance is reduced. At short distances, the potential energy increases consistently (becomes less negative or more positive) as the distance is reduced. Between these two regions there is a point at which the net force between the particles just changes from a net attraction to a net repulsion.

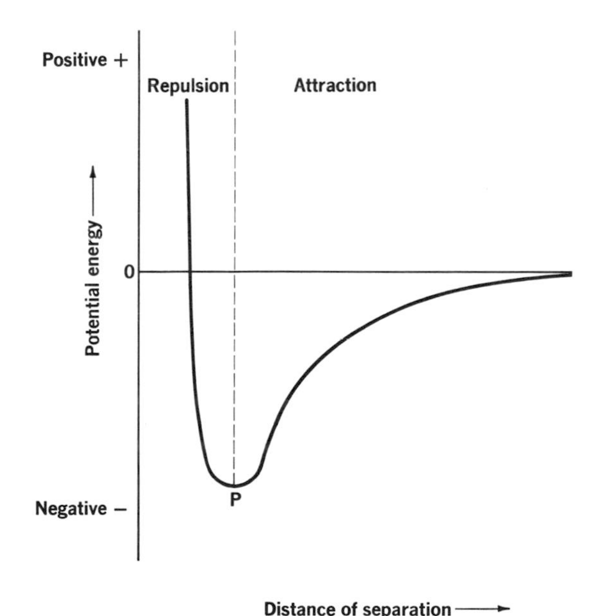

Fig. 5-3
Variation of potential energy with distance of separation. The force between the particles is one of attraction at long distances and of repulsion at short distances.

At that point (labeled *P* in Fig. 5-3) the net force is zero, and the potential energy is at a minimum.

Potential energy curves such as those shown in Figs. 5-2 and 5-3 are useful because they indicate the relative stability of the collection of particles in various configurations. A high, positive value means that the given configuration is unstable, not only relative to the isolated particles, but also relative to any other configuration in which the potential energy is lower. Thus we can imagine a series of rocks perched at various points on a steep hillside. The topmost rock has the highest potential energy and is unstable with respect to both the valley floor and the other rocks on the hillside below it.

A low negative value means that the given configuration is stable, not only relative to the isolated particles, but also relative to any other configuration in which the potential energy is higher. But even the configurations of low potential energy are still unstable with respect to any other configuration of yet lower potential energy. The only configuration that is stable with respect to all other configurations is one in which the potential energy is at an absolute minimum, for example, at point *P* in Fig. 5-3.

The kinetic molecular theory

We have seen that macroscopic objects, when in motion, possess kinetic energy, and that they possess potential energy when they interact with other macroscopic objects. The *kinetic molecular theory of matter* extends these ideas to the submicroscopic realm. This theory makes the same basic assumption as does Dalton's, namely, that matter is ultimately composed of tiny discrete particles or *molecules*. The theory further assumes that the molecules are in motion and, hence, possess kinetic energy, and that they interact with one another and, hence, possess potential energy. The quantity of kinetic and potential energy possessed by the molecules is calculated on the assumption that the same laws of behavior that apply to macroscopic objects apply here.

The idea of submicroscopic particles possessing kinetic and potential energy provides a natural explanation for the deduction that any macroscopic substance in any given state has a certain, definite amount of energy deposited in it. The energy stored in the macroscopic sample of matter is simply the sum total of all the kinetic energy and all the potential energy possessed by all the constituent molecules. In the remainder of this chapter, we shall examine that idea in more detail.

A SPECIFIC MODEL

In its simplest form, the kinetic molecular theory assumes that the molecules may be regarded as hard, structureless spheres. For each substance, the spheres are thought to have a definite and characteristic size, which never changes. When the substance expands or contracts, the spheres merely move further apart or closer together. Why should such a collection of submicroscopic spheres possess potential and kinetic energy?

Let us consider *potential energy* first. We recall that potential energy is the result of forces of attraction or repulsion. If we assume that the molecules attract one another, then the collection of molecules will certainly possess potential energy. In the case of solids and liquids, certain observable properties suggest that such an assumption is reasonable. For example, it is well known that, if a solid or liquid is placed in a container of larger volume, the solid or liquid does not expand but retains its original volume. This suggests that the submicroscopic particles attract one another, for if they did not, they would move apart and the solid or liquid would expand to fill the entire container. On the other hand, gases are known to do just that: no matter how large the container, a gas will expand and distribute itself uniformly throughout it. This suggests that if the forces acting among the molecules of the gas are attractive, the molecules are so far apart that the forces have become very weak.

For molecules to possess *kinetic energy*, we must necessarily assume that they are moving about. But are they actually moving? When we see a macroscopic object such as a beaker full of water sitting at rest on a table top, it is difficult to imagine that such a still body of water is composed of molecules in motion. Yet there is now overwhelming evidence that such is indeed the case. For example, when small but microscopically visible particles are suspended in a drop of water and examined under a microscope, the particles are seen to be moving constantly in zigzag paths. A plausible explanation of the zigzag motion is that the particles are constantly being bombarded by fast-moving but invisible water molecules.

The kinetic molecular theory postulates that the kinetic energy of the molecules is directly related to the temperature of the sample. As the temperature rises, so does the kinetic energy. Since the mass of the molecules remains constant, this means that their average velocity must be increasing as the temperature rises.

We shall now apply this model and discuss the energy relationships and several other properties of the three states of matter.

The solid state

Solids have a relatively high density and are almost incompressible. (See p. 19). These facts suggest that the molecules in a solid are closely packed and virtually at touching distances, so that it is difficult to reduce the average distance between them.

It has already been mentioned that pure solids are crystalline. For example, crystals of the substance calcite are shown in Fig. 5-4. The regular geometrical shape of crystals is perhaps the most obvious, but by no means the only, indication of a highly ordered inner structure, in which the molecules are arranged in an orderly manner. Such an orderly arrangement of molecules is known as a *lattice,* and the positions occupied by the molecules are known as *lattice sites.* For example, the crystal lattice inferred for metallic copper is shown in Fig. 5-5. The individual copper atoms are represented as little spheres, and the crystal is seen to be an orderly pile of such spheres.

Molecules in close proximity attract one another, although nothing has yet been said about the nature of the attraction. Since each molecule has a characteristic mass, we might expect some attraction due to the force of gravity. Calculation shows, however, that the gravitational attraction is fairly unimportant; the present view is that the major part of the attraction results from electrical interactions. This view is based on some experiments, to be discussed in Ch. 8, which indicate that molecules are best described not as structureless spheres, but rather, as assemblies of still smaller particles that bear electrical charges.

Fig. 5-4
Crystals of calcite ($CaCO_3$).

Now let us consider what happens to the crystal of copper when energy is added. One observes that the temperature rises and that the crystal expands slightly. The latter observation suggests either that the lattice sites become just a little larger, or that some *holes* are being created in the original lattice. A hole is a lattice site that would normally be occupied but is vacated as the atom moves to the surface of the crystal. The crystal thus expands.

According to the kinetic molecular theory, that is only part of the story, however. The rise in temperature suggests that there is also an increase in the kinetic energy of the molecules in the crystal. The type of motion that is responsible for the kinetic energy is best described as a vibration, or back-and-forth motion, of the individual molecules. A molecule confined to a lattice site is somewhat like a billiard ball confined to a box that is slightly larger than the ball. Just as some empty space will allow the ball to "rattle" in its box, so a little extra space around each molecule will allow the molecule to vibrate within the

Fig. 5-5
Lattice arrangement of the atoms in a crystal of copper [after W. Barlow, *Nature, 29,* 186 (1883)].

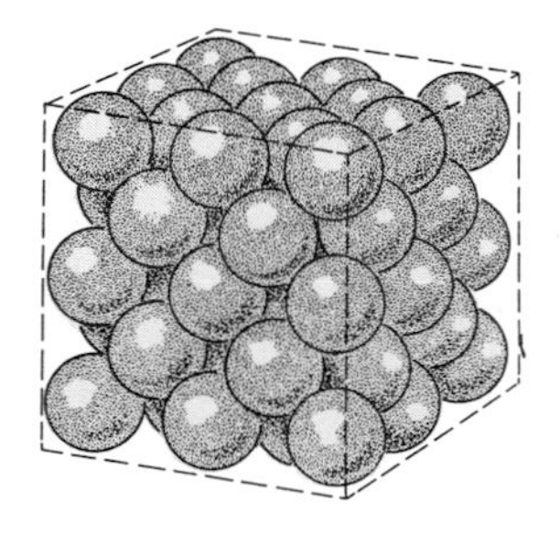

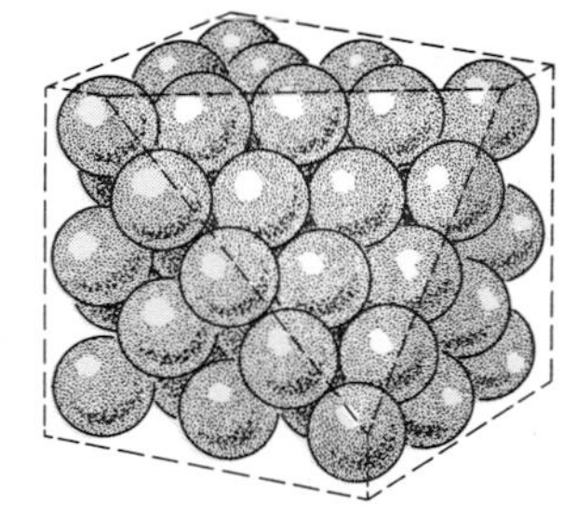

confines of its lattice site. As energy is added, that motion becomes faster and faster until the melting point is reached.

The liquid state

The macroscopic properties of liquids are intermediate between those of gases and of solids. Like solids, liquids have a high density and small compressibility; unlike solids, they are fluid, that is, they can be poured into containers of any desired shape and adapt themselves to the shape of the container.

To explain the properties of liquids, we shall adopt a model that is in accord with many experimental observations. According to this model, the immediate environment of each molecule resembles a slightly distorted crystal lattice. However, because of the cumulative effect of the distortions, the crystal-like order is quickly lost as one moves away from the given molecule. In addition, the distortions are such as to leave a fairly large fraction, perhaps several per cent, of holes that are large enough to accommodate a molecule.† These holes profoundly affect the properties of the liquid. In solids, the motion of a molecule is limited to a vibration centered on its lattice site. In liquids, molecules can vibrate in an analogous fashion, but they are also able to move through the body of the liquid. The details of that translational motion can be described as follows: owing to the motion of other molecules, the lattice site adjacent to a given molecule becomes empty. A molecule moves into the empty site, thus vacating its former site. This motion permits a second molecule to move into the newly vacated site; next a third molecule moves into the site vacated by the second one, and so on. In time, each molecule, moving from site to site, will have wandered through the entire body of the liquid.

The presence of a large number of vacant sites leaves more flexibility in the arrangement of the molecules and makes it easier to distort the liquid lattice. Hence, a liquid lattice is able to adapt its shape to that of the container. This behavior is shown pictorially in Fig. 5-6.

† Most substances expand in volume by several per cent when they melt at atmospheric pressure.

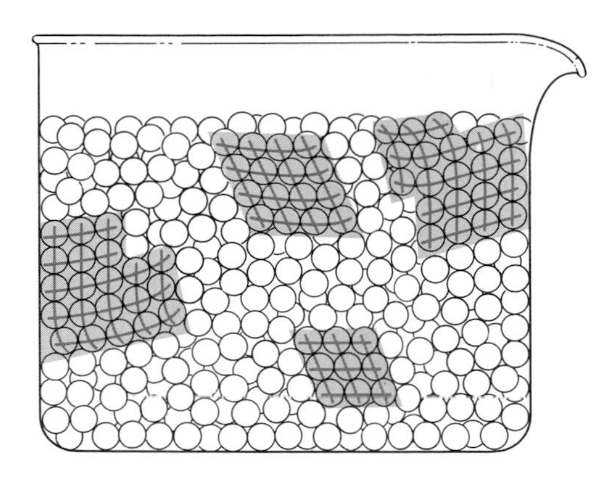

Fig. 5-6
Lattice arrangement of the molecules of a liquid.

The presence of a fairly large number of empty sites, the distortion of the lattice, and the possibility of motion of molecules from site to site, all lead to an increase in the energy of the liquid. Hence, the energy of a liquid is greater than that of the corresponding solid at the same temperature.

The gaseous state

When a liquid is converted into a gas (vapor) by boiling or evaporation, the gas occupies an enormously larger volume and possesses considerably more energy than does the liquid from which it was obtained. For example, when water is converted to steam at 100°C and one atmosphere of pressure, the volume increases 1700-fold, and 540 calories of energy must be supplied for every gram of water vaporized. That is a lot of energy; by comparison, 80 calories will melt one gram of ice, and 100 calories are required to heat one gram of water from 0° to 100°C.

Although the steam occupies a much larger volume than the water from which it was obtained, we wish to emphasize again that the volume of the individual molecules remains constant. The process is *not* like the popping of popcorn in which the individual particles increase in size. Since the total volume increases markedly, yet the volume of each individual molecule remains constant, the average distance between the molecules is greatly increased. In fact, the gaseous state is aptly described as consisting mostly of empty space, with a molecule here and there. This model explains also the high compressibility that is characteristic of gases.

How does the kinetic molecular theory explain the large increase in internal energy when a liquid is converted to a gas? The gas molecules are far apart, and we must therefore suppose that their interaction is very weak. This means that the potential energy is close to zero. On the other hand, in the liquid phase, the potential energy is quite negative since the molecules are close together and therefore attract one another strongly. In converting a liquid to a gas, those attractive forces must be overcome, and hence energy must be supplied: The potential energy increases from a negative value to virtually zero. But if the potential energy of a gas is virtually zero, then the internal energy of a gas becomes equal to the *kinetic* energy of the molecules. As a corollary, when the gas is heated, the added energy is used entirely to increase the kinetic energy of the molecules.

ADDITIONAL OBSERVATIONS CONSISTENT WITH THIS MODEL

The kinetic molecular model of the gaseous state is readily applied to other observations. For example:

1. Gases can be mixed in all proportions to form homogeneous solutions. This behavior is quite different from that of liquids or solids, which often fail to form homogeneous solutions. The behavior is readily explained, however, by a model which depicts a gas as consisting mostly of empty space, with each molecule moving independently of all others. There is always room for additional molecules. Since each molecule is free to move throughout the entire container, the impression received by an observer is that of a homogeneous solution.
2. Whenever a gas is placed in a container, one observes that a pressure is exerted on the walls of the container. This observation is readily explained by the assumption that the molecules of the gas are in constant motion. It is therefore plausible that many of the molecules are colliding with the walls of the container at any given instant. Each time a collision with the wall occurs, the molecule exerts a small force, or push, on the wall. The vast number of molecules involved gives the impression of a steady pressure.

   The same model can explain additional facts concerning gas pressure. Thus, it is known that the pressure drops when some of the gas is let out of the container, e.g., when air is let out of an automobile tire. The explanation is obvious: letting some of the gas out reduces the number of molecules, and hence, the number of collisions with the walls in unit time.

Another fact concerning gases is that the pressure exerted by a gas increases whenever the gas is heated in a closed container of fixed volume. To explain this fact, we recall that the added energy serves to increase the kinetic energy of the gas molecules. The molecules are therefore moving faster at the higher temperature. This means that it takes less time for a molecule to travel from wall to wall, so that there are more collisions per unit time. Also, since the molecules are more energetic, each collision exerts a greater force. Both factors lead to an increase in the total force exerted by all the molecules, and hence, to an increase in the observed pressure.

3. When one portion of a gas is warmer than another, heat flows away from the warmer region until the temperature is uniform. In the warm region, the molecules have a higher average kinetic energy than in the cold region. When the temperature has become uniform, the average kinetic energy of the molecules is the same throughout the vessel. This process cannot be regarded as the simple mixing of fast and slow molecules, but involves transfer of some of the energy of the fast-moving molecules to the slower ones, speeding them up.

   The moving molecules collide not only with the wall, but also with one another whenever their paths cross. The collisions then provide a mechanism for the exchange of kinetic energy between the fast- and slow-moving molecules. As a result, the original identity of the groups of molecules as "fast" or "slow" is soon lost.

Collisions between molecules and the distribution of their speeds

If we could follow the motion of a single gas molecule, we would find that it moves in uniform motion (with constant speed and direction) during the time between collisions, and that it experiences a change in motion at each collision. Sometimes the molecule will emerge from a collision with exceptionally high velocity, at other times it will become almost stationary, and most often it will emerge with some intermediate speed. If we could monitor the motion of the molecule over a sufficient period of time, we could construct a graph showing the relative frequency, or probability, that the molecule is traveling with any given speed. Alternatively, if we could record the motion of *all* the gas molecules simultaneously, we could construct a similar graph, showing the probability that a molecule chosen at random is traveling with any given speed at that particular moment. Either graph would show the distribution of molecular speeds under the given conditions.

The distribution of molecular speeds in a gas is of central importance in kinetic molecular theory, because a host of macroscopic properties of gases can be predicted from it. An exact mathematical equation for the distribution was derived about 100 years ago by a German, Ludwig Boltzmann, and a Briton, James Clerk Maxwell, by a purely theoretical approach. Their analysis showed that, when a gas is placed in a container and left alone so that no energy can enter or leave, the distribution of the speeds will soon become steady. That is to say, the percentage of molecules in any particular speed range becomes constant and remains so indefinitely. The actual numerical values of these percentages depend on the temperature and on the nature of the gas, but the distribution always has the general form of the histograms shown in Fig. 5-7.

The distribution remains steady owing to the operation of the laws of chance: there are so many molecules and so many collisions occurring all the time that when one molecule changes its speed and vacates its position in the speed distribution, there is a strong probability that some other molecule in some other collision will have changed its speed so as to take up that position in the speed distribution.

The histograms shown in Fig. 5-7 look very much like histograms showing the distribution of incomes in a typical society or of grades in a student population. Examining the upper histogram, we can read from it that at 0°C fewer than 2% of the molecules have speeds in the range between zero and 300 feet per second; 6% have speeds in the range 300 to 600 feet per second; and so on. The largest group, 18% of the molecules, has speeds in the range 1200 to 1500 feet per second, and at still higher speeds the percentages become progressively smaller.

The 1000°C histogram has a similar shape, but the maximum occurs at a higher speed, and the fraction of molecules with very high speeds is much greater. In later chapters we shall find the molecules that travel at high speeds of special interest, because they are the ones that are capable of undergoing chemical reaction.

Fig. 5-7
Histograms showing the velocity distribution for nitrogen molecules at 0°C and at 1000°C. Note that the fraction of fast-moving molecules is much greater at the higher temperature.

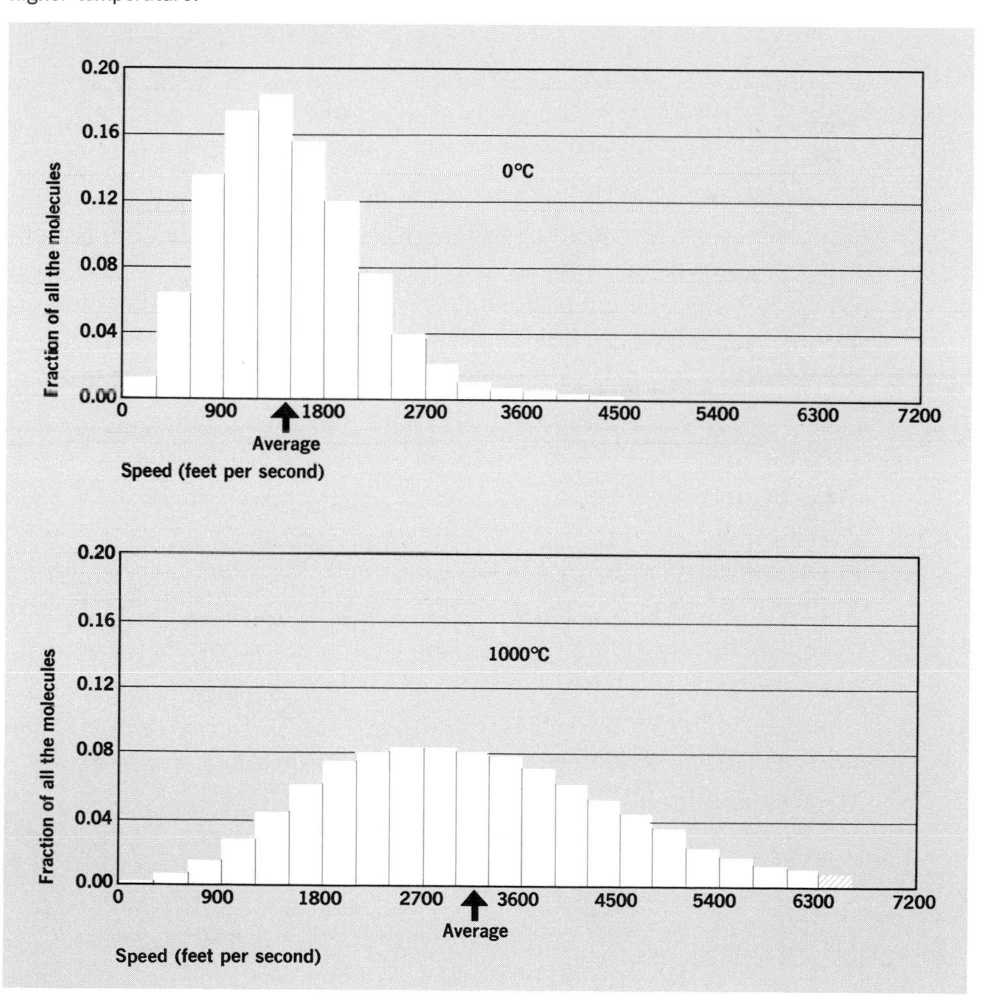

Some numerical results†

The kinetic molecular theory allows us to estimate the values of certain properties of molecules, such as size and collision frequency, that cannot be observed directly. We shall now present some of the results, without going into the details of their derivation. It is fair to say that these results are quite accurate, provided that the kinetic molecular theory itself constitutes an accurate model. They will do much to help us visualize the submicroscopic picture of the gaseous state.

The average speeds of the molecules of typical gases have been computed from the theoretical equation giving the distribution of molecular velocities for the gases in question. The results are as follows:

Hydrogen:	169,000 cm/sec at 0°	(3,770 mi/hr)
Nitrogen:	45,400 cm/sec at 0°	(1,015 mi/hr)
Carbon dioxide:	36,300 cm/sec at 0°	(811 mi/hr)

Average speeds increase slowly with increasing temperature and are about double these figures at 800°C, which is roughly the temperature inside a hot gas flame. Expressed in miles per hour, the molecules travel at speeds ranging from 800 to 4000 mi/hr, speeds comparable to those reached by the fastest jet planes.

Another quantity of interest is the size of the molecules. Some specific values for the average diameter of actual molecules are as follows:

Hydrogen molecules:	2.2×10^{-8} cm in diameter
Nitrogen molecules:	3.2×10^{-8} cm in diameter
Carbon dioxide molecules:	3.6×10^{-8} cm in diameter

Note that the average diameters of these molecules are on the order of 1/40,000,000 cm. Of course, this is much too small to be seen with the naked eye, or even with a good microscope that can magnify an object to a thousand times its original size. The smallest object that can be seen under such a microscope must be at least 1/40,000 cm in diameter, or still a thousand times larger than a typical molecule.

As the molecules traverse the container in which they are held

† If the reader is unfamiliar with exponential notation, he should review Appendix B before reading this section.

captive, their paths cross from time to time, and a collision results. It is of interest to compare the average distance a molecule will travel between collisions with the size of the apparatus. Of course, that distance depends on how many molecules there are in the container, because the frequency with which the paths will cross will increase with the number of molecules. Since the pressure of the gas is also proportional to the number of molecules, we shall tabulate the average distance between collisions as a function of the pressure.

AVERAGE DISTANCE BETWEEN COLLISIONS

	Pressure		
Gas	1 atm	$\frac{1}{1000}$ atm	$\frac{1}{1,000,000}$ atm
Hydrogen	1.12×10^{-5} cm	0.0112 cm	11.2 cm
Nitrogen	0.60×10^{-5} cm	0.0060 cm	6.0 cm
Carbon dioxide	0.40×10^{-5} cm	0.0040 cm	4.0 cm

The first column of the preceding table shows that at 1 atm pressure, the average distance between collisions is only about 1/100,000 cm, which is very small indeed. But that distance increases with decreasing pressure, and at a pressure of 1/1,000,000 atm, which is readily attainable in the laboratory with modern vacuum pumps, the molecules travel for several centimeters between collisions. Such a distance is of the same order of magnitude as the wall-to-wall distance of a small laboratory apparatus, so that many molecules will be traveling from wall to wall without colliding at all with other molecules.

Perhaps the most impressive figure is the number of collisions experienced by one molecule in 1 sec. That number again is dependent on the pressure of the gas and is listed as a function of the pressure.

COLLISIONS EXPERIENCED BY ONE MOLECULE IN 1 SEC.

	Pressure		
Gas, at 0°C	1 atm	$\frac{1}{1000}$ atm	$\frac{1}{1,000,000}$ atm
Hydrogen	1.6×10^{10}	1.6×10^{7}	16,000
Nitrogen	0.8×10^{10}	0.8×10^{7}	8,000
Carbon dioxide	0.95×10^{10}	0.95×10^{7}	9,500

At 1 atm, the number of collisions experienced by one molecule in 1 sec is about ten billion. In order to visualize the enormity of that number, imagine a prize fighter hitting a punching bag at the rate of three times a second. He will have to stand there punching continuously for 100 years before he has hit the punching bag as many times as one molecule gets hit in one second.

As the pressure is lowered, the frequency of the collisions decreases, but even at 1/1,000,000 atm, it is still about 10,000/sec. Collisions between molecules are important to an understanding of chemical reactions: It is the intimate contact during a collision that provides the opportunity for a chemical reaction to occur.

At any given temperature, the average kinetic energy of translation is exactly the same for all gases. This is true even though the weights of the molecules may vary from substance to substance, because the heavier molecules move just enough more slowly than the lighter ones to result in exact equality of the average translational energies. For example, we saw that hydrogen has a greater average velocity than nitrogen, but the hydrogen molecules are also lighter than the nitrogen molecules, and the work required to accelerate a hydrogen molecule from rest to its average velocity is exactly equal to that required to accelerate a nitrogen molecule from rest to its average velocity.

Energy changes in chemical reactions

Chemical reactions generally occur with the evolution or absorption of energy. A familiar example in which energy is liberated is the combustion of coal. This is a chemical reaction in which coal, a slightly impure form of the element carbon, reacts with oxygen in the air to produce carbon dioxide gas. An example of a reaction in which energy is absorbed is the decomposition of water into the elements, hydrogen and oxygen, by the passage of an electric current.

The theoretical treatment of energy changes in chemical reactions is necessarily more complex than that in physical changes because the molecules themselves are changing. We shall therefore postpone further discussion until we have more information about the binding of atoms in molecules.

Suggestions for further reading

Alder, B. J., and Wainwright, T. E., "Molecular Motions," *Scientific American*, October 1959, p. 113.

Ashcroft, N. W., "Liquid Metals," *Scientific American*, July 1969, p. 72.

Bernal, J. D., "The Structure of Liquids," *Scientific American*, August 1960, p. 124.

Bragg, L., "X-ray Crystallography," *Scientific American*, July 1968, p. 58.

Brenner, S. S., "Metal Whiskers," *Scientific American*, July 1960, p. 64.

Eastman, G. Y., "The Heat Pipe," *Scientific American*, May 1968, p. 38.

Lounasmaa, O. V., "New Methods for Approaching Absolute Zero," *Scientific American*, December 1969, p. 26.

Mason, E. A., and Evans, R. B., "Graham's Laws: Simple Demonstration of Gases in Motion," *Journal of Chemical Education, 46*, June 1969, p. 358.

Wannier, G. H., "The Nature of Solids," *American Scientist*, December 1952, p. 39.

Wilson, M., "Count Rumford," *Scientific American*, October 1960, p. 158.

Wolfgang, R., "Chemistry at High Velocities," *Scientific American*, January 1966, p. 82.

Questions

In questions 1–10, complete the statement by supplying the missing words or phrases.

1. A calorimeter is a device for measuring _______________ .

2. It would take _______________ calories of heat to raise the temperature of 20 g of liquid water from 30°C to 70°C.

3. If the temperature of a piece of iron is nonuniform, heat will flow from _______________ to _______________ . The heat flow stops when _______________ _______________ .

4. The translational kinetic energy of a moving body depends both on the _______________ and on the _______________ .

5. An isolated particle of matter may possess ______________ energy, but it takes at least two particles to have ______________ energy.

6. In a crystalline solid, the molecules are found to exist in an orderly arrangement called a ______________ .

7. The average translational energy of the molecules in a gas is independent of the nature of the gas and depends only on ______________ .

8. According to the kinetic molecular theory, as a liquid is converted into a gas, the potential energy of the molecules of the substance ______________ . Indeed, in the gas phase the molecules are so far apart that their potential energy is virtually ______________ .

9. The mean free path of the molecules in a gas is increased most effectively by ______________ . Thus a gas molecule will travel (on the average) several inches between collisions if ______________ (state the conditions).

10. For a gas which has been maintained at constant temperature and pressure for a long period of time, the speed of the molecules is changing continually. However, the ______________ of the molecular speeds remains the same.

11. Suppose that a mixture of ice and water exists at 0°C in a thermos bottle that provides perfect thermal insulation. What will happen if:
(*a*) The mixture of ice and water in the thermos bottle is left alone?
(*b*) The mixture of ice and water in the thermos bottle is stirred mechanically?
(*c*) The outside of the thermos bottle is tapped vigorously for several hours?
(*d*) An electrical heating wire (such as in a toaster) is immersed in the mixture and a voltage is applied?

12. For each of the following changes, indicate whether the energy of the substance increases, decreases, or stays the same.
(*a*) Water, initially at 25°C, is distilled completely and the distillate is allowed to come to 25°C.
(*b*) Water evaporates from an evaporating dish at room temperature.
(*c*) The gaseous refrigerant, Freon, is compressed and liquefies while a stream of cool air blows over the apparatus.
(*d*) Air is pumped into a bicycle tire.

13. Two particles that attract each other are separated initially by a great distance. The two particles are then brought together slowly until they just touch. Finally, the two particles are squeezed together with considerable force. Draw a diagram showing how the potential energy varies with the distance between the centers of the particles.

14. (*a*) Explain, in terms of kinetic molecular theory, what happens when the temperature of a gas is raised. How does your explanation account for the increase in pressure when the gas is heated at constant volume?

(*b*) What would the universe be like if there were absolutely no forces of interaction among molecules?

15.* A solid that is not crystalline is called *amorphous.* In a typical amorphous solid the molecules are similar to long, flexible chains that get entangled and intertwined to produce the macroscopic rigidity. Some materials, for instance polyethylene, may exist either in a crystalline state or in an amorphous state. Compare the two states at a given temperature with respect to potential energy and kinetic energy of the molecules. Which state would you expect to be the more stable?

* Questions with an asterisk (*) are more difficult or require supplementary reading.

False facts are highly injurious to the progress of science, for they often endure long; but false views, if supported by some evidence, do little harm, for everyone takes a salutary pleasure in proving their falseness; and when this is done, one path toward error is closed and the road to truth is often at the same time opened.

CHARLES DARWIN (1809–1882)

Six

MOLECULAR FORMULAS AND THE ATOMIC WEIGHT SCALE

ONCE THE concept is accepted that matter is made up of molecules, and that the molecules are, in turn, made up of atoms, it becomes necessary to return to the problem that baffled Dalton, namely, how to determine the *number* of each kind of atom in the molecule. (See page 55.) We have seen that, as long as the only information available is the percentage composition, it is not possible to solve the problem uniquely and obtain true molecular formulas. In this chapter, we shall learn how the problem was actually solved in a very direct way by a felicitous combination of the experiments of J. L. Gay-Lussac and the theories of Amadeo Avogadro.

Gay-Lussac's law of combining volumes

The key to finding correct molecular formulas lay hidden in some experiments on chemical reactions in the gaseous state, reported in 1809 by the French chemist, Gay-Lussac. These experiments are a classic example of pure research and were carried out just for the sake of gaining further knowledge, without trying to solve any specific practical or industrial problems. Yet, when we consider the practical value of Gay-Lussac's purely scientific experiments in the long run—how they provided the information that solved the problem of molecular formulas and thereby sparked the development of modern chemistry, with all the practical benefits which have resulted—it is difficult to conceive of any piece of applied research or gadget that could have been of greater practical value. The moral taught by this bit of history is worth learning. Some politicians and industrialists nowadays oppose financial support for scientific research unless it be directed towards the solution of immediate practical problems. There is no doubt that such applied research is needed—often urgently—and should be supported. But there is also no doubt that applied research is no substitute for pure research. For best results in the long run, the two kinds of research must go on side-by-side.

Gay-Lussac was concerned with the ratios of the *volumes* in which gaseous compounds react to form new compounds. He began his investigation by causing equal volumes of various gases (under the same conditions of temperature and pressure) to react with one another; he then determined the volumes of the products as well as those of unreacted starting materials (measured again at the same temperature and pressure). By subtracting the volume of unreacted gas from the initial volume, he obtained the volume that had actually reacted.

One of the reactions Gay-Lussac studied was the combination of hydrogen with oxygen to form steam. Since these relationships apply only to gaseous reactions, it is obvious that this reaction must be carried out at a temperature greater than 100 °C to prevent the steam from condensing to form liquid water. As is shown schematically in Fig. 6-1, Gay-Lussac found that when reaction took place, two volumes of hydrogen gas (at 100 °C and 1 atm pressure) reacted with one volume of oxygen gas (at 100° C and 1 atm pressure) to produce two volumes of steam (at 100 °C and 1 atm pressure).†

Or again, as shown in Fig. 6–2, two volumes of carbon monoxide

Fig. 6-1
The result of Gay-Lussac's experiment with hydrogen gas and oxygen gas.

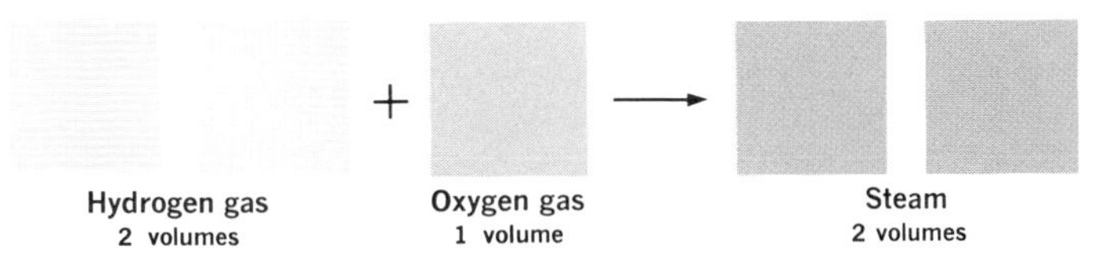

† Chemists use an arrow (⟶) to denote the formation of products in a chemical reaction.

Joseph Louis Gay-Lussac:
1778–1850
An exceedingly versatile scientist who contributed importantly to synthetic and industrial chemistry as well as to the atomic theory. [Photograph courtesy Brown Brothers.]

gas reacted with one volume of oxygen gas to produce two volumes of carbon dioxide gas.

In yet a third example (which was studied by A. Berthollet), nitrogen gas and hydrogen gas reacted to form ammonia gas (see Fig. 6-3);

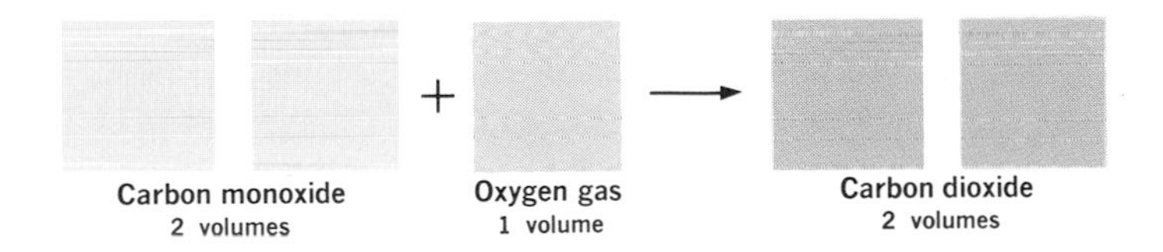

Fig. 6-2
The result of Gay-Lussac's experiment with carbon monoxide gas and oxygen gas.

Fig. 6-3
The result of Berthollet's experiment with nitrogen gas and hydrogen gas.

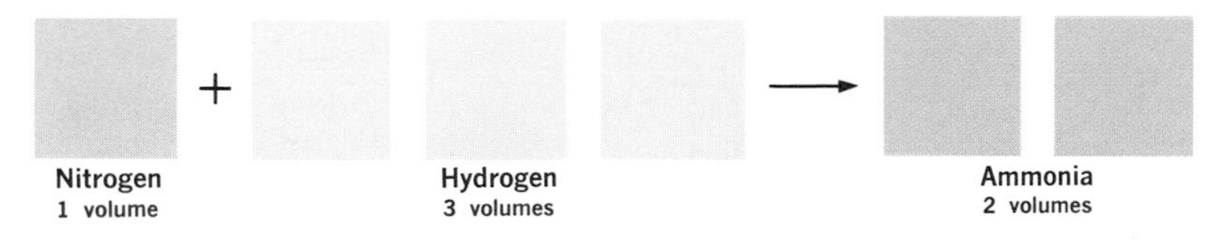

here one volume of nitrogen reacted with three volumes of hydrogen to produce two volumes of ammonia. From these and other experiments, Gay-Lussac concluded:

When gases unite to form new compounds, the volumes of the reactants and those of the products are in the ratio of small whole numbers.

This generalization is referred to as Gay-Lussac's LAW OF COMBINING VOLUMES.

On the basis of these observations, Gay-Lussac suggested that there must be a simple proportional relationship between the volumes of the reactants and the numbers of submicroscopic particles of each substance involved in the reaction. Unfortunately for the immediate acceptance of his theory, Dalton refused to take Gay-Lussac's experiments seriously because he firmly believed the molecules of *elementary* substances to be indivisible, and on that basis could not see how one volume of an *element* could yield more than one volume of a compound. For example, he could not see how one volume of oxygen could yield two volumes of steam, and therefore unjustly ascribed Gay-Lussac's observations to experimental error.

Avogadro's hypothesis

Another chemist, an Italian named Amadeo Avogadro, saw in the law of combining volumes an important clue to the molecular formulas of the reactants and products. He suggested that the problem could be solved by making the following assumptions:

1. Equal volumes of *all gases* under the same conditions of temperature and pressure contain the *same number* of molecules.
2. Molecules of certain elements (oxygen, hydrogen, nitrogen, etc.) are not indivisible, as had been supposed by Dalton, but instead consist of two identical atoms that are easily separated when the element reacts to form a compound.

Armed with these assumptions it becomes easy to deduce, for example, the correct molecular formula of water. Recall that *two* volumes of hydrogen react with *one* volume of oxygen to form *two* volumes of steam. Therefore according to Avogadro's first assumption, twice as many hydrogen molecules as oxygen molecules are required to form a given amount of water. Also, the number of water molecules that are formed is equal to the number of hydrogen molecules that are used up because the volumes of steam and hydrogen are equal.

To complete the determination of the molecular formula of water we must make use of Avogadro's second postulate. If each oxygen molecule contains two identical oxygen atoms, and each oxygen molecule produces two water molecules, then each water molecule must contain one oxygen atom. The same line of reasoning, when applied to the relation between the volumes of hydrogen and of water (2:2), leads to the conclusion that each water molecule contains *two* hydrogen atoms. Thus, the molecular formula of water must be H_2O. Diagrammatically, this can be depicted as shown in Fig. 6-4. Note that each volume contains the same *number* of molecules (in this diagram, five molecules per volume), and that the five molecules of oxygen are capable of forming ten molecules of water (2 volumes), whereas it takes ten molecules of hydrogen to form this amount of water.

Avogadro's first postulate, which states that equal volumes of gases contain equal numbers of molecules, can today be derived directly from the kinetic molecular theory. There is, therefore, general agreement nowadays that this hypothesis is justified.

Dalton objected vigorously to the second postulate, as we mentioned before. However, his objections are readily met by the assumption that certain elementary substances (the gases mentioned above) under normal circumstances form aggregates of *like* atoms in a way qualitatively no different from the aggregates of *unlike* atoms (molecules)

Fig. 6-4
Avogadro's representation of Gay-Lussac's experiment on water.

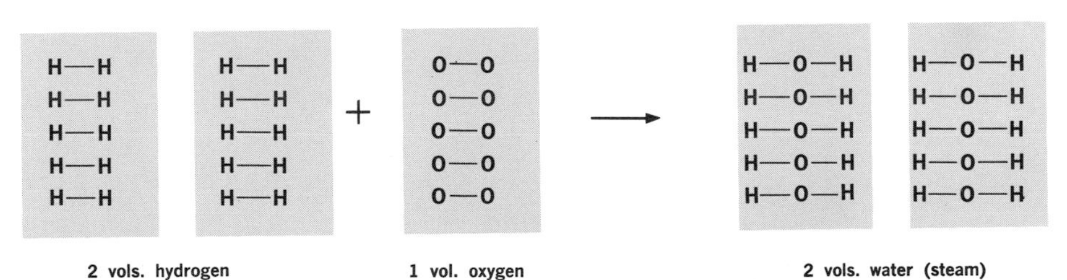

that we recognize as compounds. But why did Avogadro decide that the molecules formed by these elements contain only two atoms? Why not three or four, or some other number? The answer can be found only by a careful examination of Gay-Lussac's data. In all of the experiments involving oxygen, hydrogen, or nitrogen, no one has ever observed that one volume of one of these gases produces more than two volumes of product. This suggests that there are two atoms per molecule. If the molecules of an element X were in fact composed of *n* identical atoms, one would expect to find at least a few reactions in which one volume of gaseous X yields *n* volumes of a gaseous product, that is, in which the product molecules contain only one X-atom per molecule.

Later studies have resulted in the discovery that there are indeed some elements with molecules that contain more than two atoms. In the case of phosphorus, which contains four atoms per molecule, one obtains a maximum of four volumes of product from one volume of phosphorus vapor. For example, phosphorus reacts with hydrogen to form phosphine with the volume relationships shown in Fig. 6-5. This

Fig. 6-5
Phosphorus reacts with hydrogen to form phosphine.

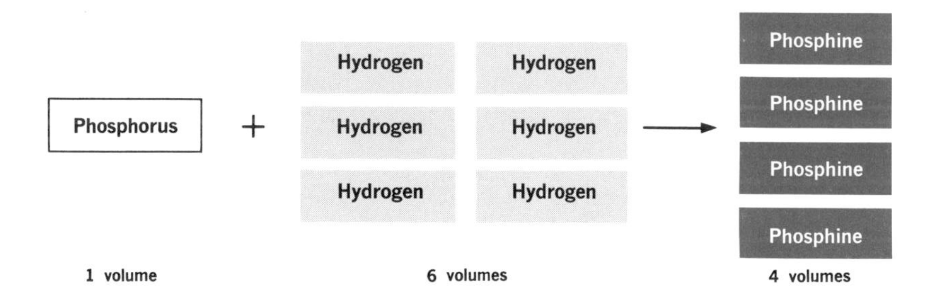

reaction not only shows that phosphorus contains four atoms per molecule, but it also enables us to deduce the molecular formula of phosphine if we assume that hydrogen molecules consist of two identical atoms. Since six volumes of hydrogen are required to produce four volumes of phosphine, six hydrogen molecules, or twelve hydrogen atoms, are required to produce four molecules of phosphine. Then, in each phosphine molecule there must be 12/4 or 3 hydrogen atoms. The

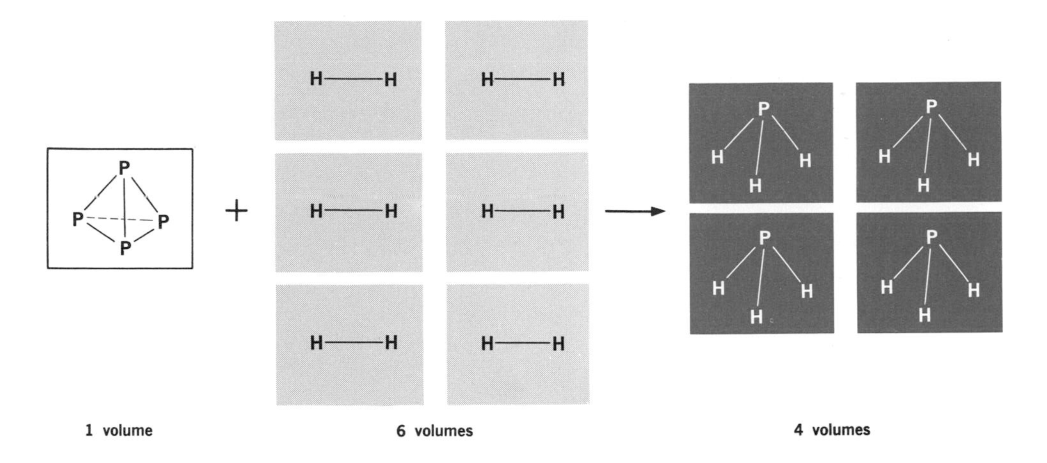

Fig. 6-6
Molecular representation of the formation of phosphine.

accepted molecular formula of phosphine is, therefore, PH_3. These relationships are shown explicitly in Fig. 6-6.

In summary, by accepting Avogadro's hypotheses we have a direct experimental method for determining the molecular formula of any element or compound that can be obtained in the gaseous state.

Atomic weights from molecular formulas and percentage compositions of compounds

Recall that Dalton's atomic theory was based mostly on weight relationships and that it assumed a characteristic weight for the atoms of each element. The assignment of atomic weights to the individual elements was a central problem of the theory and could not be solved uniquely until molecular formulas became known. With the acceptance of Avogadro's hypotheses, molecular formulas became available, and therefore it became possible to develop an accurate scale of atomic weights.

First let us see how a knowledge of the molecular formula and the composition of a compound enables us to calculate the relative weights of the constituent atoms. For example, the molecular formula of water has been found to be H_2O, and the composition is 88.8 weight per cent (wt. %) of oxygen, and 11.2 weight per cent of hydrogen. The calculation then proceeds as follows:

$$\begin{aligned}\text{Wt. \% of hydrogen} &= \text{Wt. of hydrogen in 100 g of water} \quad (1)\\ &= \text{Number of hydrogen atoms in 100 g of water} \times \text{Wt. of 1 hydrogen atom}\\ \text{Wt. \% of oxygen} &= \text{Wt. of oxygen in 100 g of water} \quad (2)\\ &= \text{Number of oxygen atoms in 100 g of water} \times \text{Wt. of 1 oxygen atom}\end{aligned}$$

Dividing (1) by (2) gives:

$$\frac{\text{Wt. \% of hydrogen}}{\text{Wt. \% of oxygen}} = \frac{\text{Number of hydrogen atoms in 100 g of water}}{\text{Number of oxygen atoms in 100 g of water}} \times \frac{\text{Wt. of 1 hydrogen atom}}{\text{Wt. of 1 oxygen atom}} \quad (3)$$

The quantity

$$\frac{\text{Wt. of 1 hydrogen atom}}{\text{Wt. of 1 oxygen atom}}$$

is precisely the ratio of atomic weights we are looking for. It can be calculated because the percentage composition is known, and because the quantity

$$\frac{\text{Number of hydrogen atoms in 100 g of water}}{\text{Number of oxygen atoms in 100 g of water}}$$

must equal the quantity

$$\frac{\text{Number of hydrogen atoms in 1 molecule}}{\text{Number of oxygen atoms in 1 molecule}}$$

The latter is equal to $2/1$ because the molecular formula is H_2O. Thus,

$$\frac{11.2}{88.8} = \frac{2}{1} \times \frac{\text{Wt. of 1 hydrogen atom}}{\text{Wt. of 1 oxygen atom}} \quad (4)$$

Solving equation (4), we find the desired ratio of the weights of the atoms to be $11.2/2(88.8)$, or very nearly $1/16$. In other words, the weight of one hydrogen atom is very nearly one-sixteenth the weight of one oxygen atom.

The atomic weight scale

By a series of calculations of this type, involving compounds whose composition and molecular formula are known, it becomes possible to relate the weights of all of the atoms to each other. Thus, we are able to set up an *atomic weight scale,* that is, a table that lists the weights of all the elements relative to some common standard. Until quite recently, the standard universally chosen was the element oxygen. Oxygen was chosen for practical reasons because most of the other elements form compounds with it. The weights of the other elements were then compared to oxygen. In this comparison, oxygen is assigned the atomic weight of exactly 16. This assignment makes the atomic weight of the lightest element, hydrogen, very nearly 1.0, and therefore avoids the use of atomic weights less than one. In 1961, the atomic weight standard was changed by the International Union of Pure and Applied Chemistry,† and the atomic weight of oxygen is now taken as 15.9994. However, the effect of this change is quantitatively very small, and the old familiar number, 16.00, is still sufficiently accurate for most calculations.

Direct determination of atomic weights

The relative atomic weights of the gaseous elements (e.g., hydrogen, nitrogen, and chlorine) can be established directly by comparison of equal volumes of the gas and oxygen, the standard. Of course, the number of atoms in the molecules must first have been reasoned from the maximum number of volumes of gaseous product obtainable from each element. Let us suppose that a certain volume of oxygen (under given conditions) weighs exactly 10.0 g. Then an equal volume of hydrogen, which according to Avogadro contains an equal number of molecules, will be found to weigh 0.625 g. The ratio of the weights of an oxygen molecule to that of a hydrogen molecule must therefore be 10.0 : 0.625, or 16 : 1. Since each oxygen molecule contains two identical oxygen atoms, and each hydrogen molecule contains two identical hydrogen atoms, the ratio of the weight of an oxygen atom to that of a hydrogen atom must also be 16 : 1. That is to say, hydrogen will have the

† The new atomic weight standard is carbon-12. See page 98.

value of 1 on the atomic weight scale. (By using more precise weights, it has been found that the value actually is closer to 1.008.) The experiment can be repeated for the other gaseous elements and a relative weight, the atomic weight, determined for each.

Using these atomic weights, it is possible to calculate directly the relative molecular weights of various compounds on this same scale. This *molecular weight* is simply the sum of the atomic weights of the constituent elements. For example, the molecular weight of water (H_2O) is equal to two times the atomic weight of hydrogen plus the atomic weight of oxygen:

$$\underset{2 \times \text{At. wt. of H}}{2 \times 1.008} + \underset{\text{At. wt. of O}}{16.00} = \underset{\text{Mol. wt. of } H_2O}{18.016}$$

The molecular weight of nitrous oxide (N_2O) is calculated analogously from the atomic weights of oxygen and of nitrogen (14.01).

$$\underset{2 \times \text{At. wt. of N}}{2 \times 14.01} + \underset{\text{At. wt. of O}}{16.00} = \underset{\text{Mol. wt. of } N_2O}{44.02}$$

We can also determine the molecular weights of these two compounds experimentally, by comparing the weight of a volume of the compound (in the gaseous state) with an equal volume of oxygen under the same conditions of temperature and pressure. In this comparison, the molecular weight of oxygen is taken as 32, since there are two oxygen atoms (2×16.00) in each molecule of oxygen.

Atomic weights of nonvolatile elements

The atomic weights of elements that are not gaseous at reasonable temperatures and pressures obviously cannot be determined directly by the comparison of weights of gas volumes. The atomic weights of such elements may be determined by a somewhat indirect method, but one that utilizes exactly the same principle, that is, the measuring of the molecular weights and percentage compositions of gaseous *compounds* of the element in question. Let us consider the element carbon as an example. The relative weight of carbon cannot be determined directly because this element does not vaporize appreciably at temperatures below 1000 °C. But carbon does occur in a large number of gaseous compounds whose molecular weights can be determined by comparison of the weight of a known volume with that of an identical volume of oxygen. Some experimental data are tabulated below:

DATA FOR CARBON COMPOUNDS

Name of the compound	Molecular weight of the compound	Weight per cent carbon	Weight of the carbon atoms in one molecule
Carbon monoxide	28	43.0	$\frac{43}{100} \times 28 = 12$
Carbon dioxide	44	27.3	12
Methane	16	75.0	12
Acetylene	26	91.0	24
Ethylene	28	86.0	24
Propane	44	82.0	36

Examination of the last column of the table shows that the contribution made by carbon to the total molecular weights of these compounds is always 12 or some integral multiple of 12. This suggests that the relative weight of carbon, on a scale where the oxygen atom has a relative weight of 16, is 12. This conclusion is, of course, based on the assumption that in those compounds where the contribution due to carbon is 12, there is only one carbon atom.

We can conclude that this is a valid assumption only after we have carried out experiments on a large number of carbon compounds without ever finding a case in which the contribution due to carbon is less than 12.

Gram atoms, moles, and the gram molecular volume

The basic assumption in Dalton's atomic theory is that when atoms combine to form molecules, they do so in the ratio of small whole numbers. In carrying out a reaction, it is therefore convenient to weigh the elements in such quantities that the numbers of atoms that are weighed are in the same ratio as in the final compound. Thus, if we wish to prepare water, we would like to be able to weigh twice as many hydrogen atoms as oxygen atoms. It would obviously not do to weigh 2 g of hydrogen and 1 g of oxygen, since the weight of a hydrogen atom is not the same as that of an oxygen atom.

TABLE OF RELATIVE ATOMIC WEIGHTS 1969

Based on the Atomic Mass of $C^{12} = 12$

Name	Symbol	Atomic number	Atomic weight	Name	Symbol	Atomic number	Atomic weight
Actinium	Ac	89	....	Mercury	Hg	80	200.59
Aluminum	Al	13	26.9815	Molybdenum	Mo	42	95.94
Americium	Am	95	....	Neodymium	Nd	60	144.24
Antimony	Sb	51	121.75	Neon	Ne	10	20.179
Argon	Ar	18	39.948	Neptunium	Np	93	....
Arsenic	As	33	74.9216	Nickel	Ni	28	58.71
Astatine	At	85	....	Niobium	Nb	41	92.9064
Barium	Ba	56	137.34	Nitrogen	N	7	14.0067
Berkelium	Bk	97	....	Nobelium	No	102	....
Beryllium	Be	4	9.01218	Osmium	Os	76	190.2
Bismuth	Bi	83	208.9806	Oxygen	O	8	15.9994
Boron	B	5	10.81	Palladium	Pd	46	106.4
Bromine	Br	35	79.904	Phosphorus	P	15	30.9738
Cadmium	Cd	48	112.40	Platinum	Pt	78	195.09
Calcium	Ca	20	40.08	Plutonium	Pu	94	....
Californium	Cf	98	....	Polonium	Po	84	....
Carbon	C	6	12.011	Potassium	K	19	39.102
Cerium	Ce	58	140.12	Praseodymium	Pr	59	140.9077
Cesium	Cs	55	132.9055	Promethium	Pm	61	....
Chlorine	Cl	17	35.453	Protactinium	Pa	91	231.0359
Chromium	Cr	24	51.996	Radium	Ra	88	226.0254
Cobalt	Co	27	58.9332	Radon	Rn	86	....
Copper	Cu	29	63.546	Rhenium	Re	75	186.2
Curium	Cm	96	....	Rhodium	Rh	45	102.9055
Dysprosium	Dy	66	162.50	Rubidium	Rb	37	85.4678
Einsteinium	Es	99	....	Ruthenium	Ru	44	101.07
Erbium	Er	68	167.26	Samarium	Sm	62	150.4
Europium	Eu	63	151.96	Scandium	Sc	21	44.9559
Fermium	Fm	100	....	Selenium	Se	34	78.96
Fluorine	F	9	18.9984	Silicon	Si	14	28.086
Francium	Fr	87	....	Silver	Ag	47	107.868
Gadolinium	Gd	64	157.25	Sodium	Na	11	22.9898
Gallium	Ga	31	69.72	Strontium	Sr	38	87.62
Germanium	Ge	32	72.59	Sulfur	S	16	32.06
Gold	Au	79	196.9665	Tantalum	Ta	73	180.9479
Hafnium	Hf	72	178.49	Technetium	Tc	43	....
Helium	He	2	4.0026	Tellurium	Te	52	127.60
Holmium	Ho	67	164.930	Terbium	Tb	65	158.9254
Hydrogen	H	1	1.0080	Thallium	Tl	81	204.37
Indium	In	49	114.82	Thorium	Th	90	232.038
Iodine	I	53	126.9045	Thulium	Tm	69	168.9342
Iridium	Ir	77	192.22	Tin	Sn	50	118.69
Iron	Fe	26	55.847	Titanium	Ti	22	47.90
Krypton	Kr	36	83.80	Tungsten	W	74	183.85
Lanthanum	La	57	138.9055	Uranium	U	92	238.029
Lawrencium	Lw	103	....	Vanadium	V	23	50.9414
Lead	Pb	82	207.2	Xenon	Xe	54	131.30
Lithium	Li	3	6.94	Ytterbium	Yb	70	173.04
Lutetium	Lu	71	174.97	Yttrium	Y	39	88.9059
Magnesium	Mg	12	24.305	Zinc	Zn	30	65.37
Manganese	Mn	25	54.9380	Zirconium	Zr	40	91.22
Mendelevium	Md	101	....				

To ensure that we are weighing the correct numbers of each, the atomic weights of the elements must be taken into consideration. It is therefore convenient to introduce the concept of the *gram atomic weight* (or *gram atom,* for short).

One gram atomic weight (or gram atom) is that weight of the element, in grams, that is numerically equal to the atomic weight of the element.

A number of examples are tabulated below:

SOME GRAM ATOMIC WEIGHTS

Element	Atomic weight	Value of one gram atomic weight
Oxygen	16.00	16.00 g
Hydrogen	1.008	1.008 g
Sulfur	32.06	32.06 g
Carbon	12.01	12.01 g

The gram atomic weight is such a convenient unit because of the following reason:

One gram atomic weight of any element contains exactly the same number of atoms as one gram atomic weight of any other element.

A *gram molecular weight* (or *mole*) of a pure substance is defined in an analogous manner:

One gram molecular weight (or mole) of any substance is that weight of the substance, in grams, that is numerically equal to the molecular weight of the substance.

It can be shown that one gram molecular weight of any substance contains exactly the same number of molecules as one gram molecular weight of any other substance.

Examples of gram molecular weights for several substances are given below:

SOME GRAM MOLECULAR WEIGHTS

Substance	Molecular weight	Value of one gram molecular weight
Oxygen	32.00	32.00 g
Water	18.0	18.0 g
Carbon dioxide	44.0	44.0 g
Methane	16.0	16.0 g

The definition of gram molecular weight leads to another unit that is very useful in chemical calculations. This unit is the *gram molecular volume.*

The gram molecular volume is the volume occupied by one gram molecular weight of the given gaseous element or compound under the standard conditions of 0°C and one atmosphere pressure.

Thus, one gram molecular volume of oxygen is the volume occupied by 32.00 g of oxygen gas at 0 °C and one atmosphere pressure; this volume is found to be 22.4 liters (22,400 cc). Similarly, one gram molecular volume of carbon dioxide is the volume occupied by 44.0 g of carbon dioxide gas at 0 °C and one atmosphere pressure, and this volume turns out also to be 22.4 liters. In fact, the gram molecular volume of all *gaseous* substances is the same, namely, 22.4 liters. This should come as no surprise if we consider Avogadro's first hypothesis, that equal volumes of gases under the same conditions contain equal numbers of molecules. If we take equal numbers of carbon dioxide and oxygen molecules, the ratio of the weights of the two samples must be the same as the ratio of their molecular weights. Conversely, if we take amounts of two gaseous substances in ratio to their molecular weights, their volumes must be equal.

A further corollary exists. If we take one gram atomic weight or gram molecular weight of any element or compound, whether solid, liquid, or gaseous, it will contain a specific number of atoms or molecules, which will be *the same for all substances.* This number is called Avogadro's number. Avogadro's number is fundamentally important whenever the properties of individual molecules are studied.

There are several rather indirect methods for determining Avogadro's number. All of them give concordant answers, namely, 6.02×10^{23} atoms in one gram atomic weight, or molecules in one gram molecular weight. The number, 6.02×10^{23}, represents an unimaginably large number of particles. Thus, if all the molecules in one gram molecular weight of water (18 g, about one tablespoon) were enlarged to the size of sand grains, they would form a pile of sand over one mile high and one mile on each side. The immensity of the number is suggested also by the fact that if one cup (about 13 gram molecular weights) of water were thrown into the ocean and thoroughly mixed with all the waters of all the oceans of the world, *every* cup of water subsequently removed from the ocean would contain about 1000 molecules of water from that original cupful.

In the current parlance of chemistry, the terms *gram atom* and *mole* are used not only as abbreviations for gram atomic weight and gram molecular weight (as defined above), but also as convenient synonyms for Avogadro's number. Thus the phrase "one gram atom of sulfur" is equivalent in meaning to "Avogadro's number of sulfur atoms." And the phrase "one mole of methane" or "one mole of methane molecules" is equivalent to "Avogadro's number of methane molecules."

WEIGHT OF INDIVIDUAL MOLECULES

When one knows the actual number of molecules in a given sample of material, it becomes easy to calculate the *actual* weight of a single molecule. Thus, if we take 18.0 g of water, which is one gram molecular weight, we have 6.02×10^{23} molecules. Therefore, one molecule will weigh $18.0 \div (6.02 \times 10^{23})$ g, which is equal to about

3×10^{-23} g (0.0000000000000 0000000003 g)

Using Avogadro's number, we can estimate also the dimensions of a water molecule. Since 18.0 g of liquid water, containing 6.02×10^{-23} molecules, occupies 18.0 cubic centimeters, each molecule must occupy approximately $18.0 \div (6.02 \times 10^{23})$ cc, or 3×10^{-23} cc. Assuming that the water molecules are closely packed and that each water molecule is a cube whose volume is equal to 3×10^{-23} cc, then the length of a water molecule can be computed from the formula:

$$\text{Length} = \sqrt[3]{\text{Volume}}$$

On that basis, the length of a water molecule is found to be 3.1×10^{-8} cm. Is it no wonder that we are unable to see or weigh individual molecules!

Suggestions for further reading

Faraday, M., *The Chemical History of a Candle* (New York: E. P. Dutton & Co., Inc., 1874).

Greenwood, N. N., "New International Table of Atomic Weights," *Chemistry in Britain*, April 1970, p. 119.

Lowry, T. M., *Historical Introduction to Chemistry* (London: Macmillan & Co. Ltd., 1936).

Nash, L. K., "The Atomic-Molecular Theory," *Harvard Case Histories in Experimental Science*, ed. J. B. Conant (Cambridge, Mass.: Harvard University Press, 1950), Case 4.

Tilden, W., *Famous Chemists* (New York: E. P. Dutton & Co., Inc., 1921), Chs. 9, 13.

Wichers, E., "Report of the International Commission on Atomic Weights—1961," *Journal of the American Chemical Society, 84* (1962), p. 4175. (This report changes the atomic weight scale to the basis that $C^{12} = 12$ and reviews the previous literature on atomic weights.)

Questions

1. 1.00 liter of oxygen gas weighs 1.43 g. 1.00 liter of the gaseous element argon weighs 1.78 g. (Both gas volumes are measured at 0°C and 1 atm.)
(*a*) What is the molecular weight of argon?
(*b*) If the argon molecules consist of one atom each, what is the atomic weight of argon?

2. 7.00 g of an element (which we shall denote by the symbol X) reacts with excess oxygen and is completely converted to a gaseous compound, X_2O; 5.6 liters of pure X_2O (at 0°C and 1 atm) is obtained.
(*a*) Find the atomic weight of X.
(*b*) Find the molecular weight of X_2O.

3. Using the Table of Atomic Weights, predict molecular weights for the following compounds:
(*a*) Arsenious acid, H_3AsO_3
(*b*) Nickel hydroxide, $Ni(OH)_2$ or NiO_2H_2
(*c*) Sodium nitrate, $NaNO_3$
(*d*) Calcium nitrate, $Ca(NO_3)_2$

4.* Zirconium oxide has the following composition: zirconium, 74.0 wt. %; oxygen, 26.0 wt. %.
(*a*) What is the ratio of zirconium atoms to oxygen atoms in each molecule of zirconium oxide?
(*b*) Is it possible, with the information given, to deduce the molecular weight of zirconium oxide? If not, why not?

* Questions with an asterisk (*) are more difficult or require supplementary reading.

5.* A compound of carbon and sulfur has the following composition: carbon, 15.8 wt. %; sulfur, 84.2 wt. %.
(*a*) Find the ratio of carbon atoms to sulfur atoms in each molecule of the compound.
(*b*) How many molecules are there in one gram molecular weight of the compound?

6. Find the atomic weight of an element (which we shall denote by the symbol Y) given the following information: Y reacts with fluorine to form a compound, YF_6, whose composition is: fluorine, 59.1 wt. %, and Y, 40.9 wt. %.

7. 1.00 liter of a gaseous compound of sulfur and oxygen was found to weigh 2.86 g (at 0°C and 1 atm).
(*a*) Find the molecular weight of the compound.
(*b*) Find the molecular formula of the compound.
Hint: Predict the molecular weights of possible sulfur compounds, such as SO, S_2O, SO_2, SO_3, . . . You will find that one, and only one, molecular formula is consistent with the result obtained in (*a*).

8. (*a*) What is the weight (in grams) of one molecule of carbon tetrachloride, CCl_4?
(*b*) How many CCl_4 molecules are there in one drop (0.10 g) of liquid carbon tetrachloride?

9.* (*a*) One volume of hydrogen gas reacts with excess sulfur to produce one volume of hydrogen sulfide gas. (Both volumes are measured at 0°C and 1 atm.) Apply Avogadro's hypothesis to find the number of hydrogen atoms in one molecule of hydrogen sulfide. Does this reaction give information about the number of sulfur atoms in one molecule of hydrogen sulfide?
(*b*) One volume of hydrogen sulfide gas reacts with excess oxygen to produce one volume of sulfur dioxide gas, plus an undetermined amount of water. The molecular formula of sulfur dioxide is known to be SO_2. Apply Avogadro's hypothesis to find the number of sulfur atoms in one molecule of hydrogen sulfide. Combine the result with the result obtained in (*a*) to arrive at the molecular formula for hydrogen sulfide.

10.* A chemist, investigating the composition of a species of hillside plants, isolated a pure low-boiling liquid substance. Further examination showed that substance to be a compound of carbon, hydrogen and oxygen with the following composition: carbon, 62.0 wt. %; hydrogen, 10.4 wt. %; oxygen, 27.5 wt. %. One liter of the substance in the gas phase (measured above the boiling point and corrected to 0°C and 1 atm) weighed 2.59 g. Find the molecular formula of the substance.

11.* One mole of compound **A,** which consists of 40.0 wt. % carbon, 53.3 wt. % oxygen and 6.7 wt. % hydrogen, decomposes on heating to give 1 mole of carbon dioxide (CO_2) and a compound **B** that consists entirely of carbon and hydrogen. Find the molecular formula of compound **A.**

The philosopher should be a man
willing to listen to every suggestion,
but determined to judge for himself.
He should not be biased by appearances,
have no favorite hypothesis, be of no
school, and in doctrine have no master.
Truth should be his primary object.
If to these qualities be added industry,
he may indeed hope to walk within
the veil of the temple of Nature.

MICHAEL FARADAY (1791–1867)

Seven

STOICHIOMETRY

STOICHIOMETRY IS the branch of chemistry that makes use of atomic weights and molecular formulas to calculate weight relationships in chemical reactions. When a chemist writes a chemical formula such as H_2O, he means either a water molecule, or a mole (Avogadro's number) of water molecules, or a gram molecular weight (18.016 g) of water; the precise meaning must be inferred from the context of the discussion. Or, when he writes the formula for carbon, C, he means either a carbon atom, or a gram atom of carbon, or a gram atomic weight (12.01 g) of carbon.

By the same token, when a chemist writes a chemical equation, he may be describing the reaction in terms of either molecules, or moles, or gram molecular weights (and hence grams). Thus, when he writes Eq. (1) for the formation of carbon dioxide from the elements,

$$C + O_2 \rightarrow CO_2 \tag{1}$$

he may mean any of the following:

MEANING OF THE EQUATION, $C + O_2 \rightarrow CO_2$

Level of the discussion	Meaning implied by the equation
1. Atoms and molecules (submicroscopic)	1. One carbon atom reacts with one oxygen (O_2) molecule to produce one carbon dioxide (CO_2) molecule.
2. Gram atoms and moles (macroscopic)	2. One gram atom of carbon reacts with one mole of oxygen (O_2) to produce one mole of carbon dioxide (CO_2).
3. Weight (in grams) (macroscopic)	3. 12 g (one gram-atomic weight) of carbon reacts with 32 g (one gram-molecular weight) of oxygen to produce 44 g (one gram molecular weight) of carbon dioxide.

In the above table, the first meaning, in terms of atoms and molecules, is on the submicroscopic level of Dalton's atomic theory. On the other hand, the second meaning, in terms of gram atoms and moles, and the third meaning, in terms of weight (in grams), are on the macroscopic level. The transfer in meaning from the submicroscopic to the macroscopic level is possible as a consequence of Avogadro's hypothesis: if *one* carbon atom reacts with *one* oxygen molecule, then *two* carbon atoms will react with *two* oxygen molecules, and *Avogadro's number* (6.02×10^{23}) of carbon atoms will react with *Avogadro's number* of oxygen molecules. Avogadro's number of atoms, in turn, is equal to one gram atom, and Avogadro's number of molecules is equal to one mole. Finally, the conversions from gram atoms or moles to actual weights (in grams) follow directly from the definitions of the gram atom or mole, as given in the preceding chapter (p. 99).

Molar ratios of reactants and products

Besides the specific meanings indicated above, the chemical equation also has a more general meaning, in which it tells us the *ratios* in which the substances react. Thus, Eq. (1) tells us not only that one C atom reacts with one O_2 molecule to produce one CO_2 molecule but also, in general, that the number of C atoms that react is equal to the number of O_2 molecules that react, and to the number of CO_2 molecules that form. In view of Avogadro's hypothesis, the equation also implies, on the macroscopic level, that the number of gram atoms of carbon that react is equal to the number of moles of oxygen that react, and to the number of moles of CO_2 that form. In other words, the following *ratios* are all one-to-one.

$$\frac{\text{Gram atoms of C that react}}{\text{Moles of } O_2 \text{ that react}} = \frac{1}{1}$$

$$\frac{\text{Moles of } CO_2 \text{ that form}}{\text{Moles of } O_2 \text{ that react}} = \frac{1}{1}$$

Even when the molar ratios are not simply one-to-one, they can be predicted easily from the chemical equation. For example, it was shown in the preceding chapter that Fig. 6–4 correctly depicts the formation of water molecules from the diatomic molecules of hydrogen and oxygen. In that reaction, two molecules of H_2 react with one molecule of O_2 to produce two molecules of H_2O. Using chemical shorthand, that reaction is represented conveniently by Eq. (2).

$$2\ H_2 + O_2 \rightarrow 2\ H_2O \qquad (2)$$

Note that Eq. (2) contains two distinctly different kinds of integers: those that appear as coefficients before a molecular formula, and those that appear as subscripts following the symbols for an element. These integers are interpreted as follows: a coefficient before a molecular formula indicates the number of molecules that react; a subscript following the symbol for an element indicates the number of atoms of that element in one molecule of the given substance. If no integer appears explicitly, one assumes that the coefficient or subscript is unity. Thus, in Eq. (2), the symbol "2 H_2" is shorthand for "two hydrogen molecules; and each hydrogen molecule consists of two hydrogen atoms." The symbol "O_2" is shorthand for "one oxygen molecule, which consists of two oxygen atoms." And the symbol "2 H_2O" is shorthand for "two water molecules; and each water molecule consists of two hydrogen atoms and one oxygen atom."

When Eq. (2) is interpreted on the macroscopic level, it informs us that 2 moles of H_2 react with one mole of O_2 to produce 2 moles of H_2O. The molar ratios are therefore as follows:

$$\frac{\text{Moles of } H_2 \text{ that react}}{\text{Moles of } O_2 \text{ that react}} = \frac{2}{1} \tag{3}$$

$$\frac{\text{Moles of } H_2O \text{ that form}}{\text{Moles of } H_2 \text{ that react}} = \frac{2}{2} = \frac{1}{1} \tag{4}$$

$$\frac{\text{Moles of } H_2O \text{ that form}}{\text{Moles of } O_2 \text{ that react}} = \frac{2}{1} \tag{5}$$

Note that in each case the molar ratio is identical to the ratio of coefficients in Eq. (2). The example illustrates the following general rule:

The molar ratios in which substances disappear and are produced in a chemical reaction are identical to the ratios of their coefficients in the chemical equation for the reaction.

It is understood, in applying this rule, that the chemical equation is a *balanced* chemical equation; that is, the equation is written so that there is conservation of atoms for each element. Procedures for balancing a chemical equation will be described later in this chapter.

Use of molar ratios

Because of the ease with which molar ratios can be ascertained from the chemical equation, they usually form the starting point in stoichiometric calculations.

The questions one asks in stoichiometry are macroscopic questions: How much of reactant **B** is required to react with a given quantity of **A**? And how much product can be obtained? These questions can be answered most readily if the amounts of the substances are expressed in moles. Because the mole is a macroscopic unit, it is perfectly logical to talk about such things as half a mole, or 1.581 moles, that is, quantities that are not exact integral multiples of one mole. Thus, 1.581 moles of hydrogen simply refers to an amount of hydrogen containing 1.581 x 6.02 x 10^{23} H_2 molecules, where 6.02 x 10^{23} is, of course, Avogadro's number.

To illustrate the use of molar ratios in stoichiometry, a few representative problems are discussed below.

Problem 1. Suppose you have 0.25 mole of oxygen (O_2). (a) How many moles of hydrogen (H_2) are required to react with it? (b) How many moles of water (H_2O) can be produced?

Given: The chemical equation for the reaction is given in Eq. (2). The molar ratios of the reactants and products are given in Eqs. (3) to (5).

Solution: (a) Let the unknown number of moles of hydrogen be denoted by x. Then, from Eq. (3),

$$\frac{x}{0.25} = \frac{2}{1}.$$

Therefore, x = 0.25 × 2 = 0.50 mole. Thus, 0.50 mole of hydrogen is required to react with 0.25 mole of oxygen.

(b) Let the unknown number of moles of water be denoted by y. Then, from Eq. (5),

$$\frac{y}{0.25} = \frac{2}{1}.$$

Therefore, y = 0.25 × 2 = 0.50 mole. Thus, 0.50 mole of water can be produced from 0.25 mole of oxygen.

Note that in answering the questions of Problem 1, we let x (or y) denote the unknown number of moles. Then we form the ratio of what-we-want over what-we-have, and equate that ratio to the molar ratio, as deduced from the chemical equation.

If the chemical equation shows atoms rather than molecules, then we must work with gram atoms rather than moles, but the method is analogous. For an example, let us consider the reaction of the elements lead (Pb) and sulfur (S), which has been described previously (p. 45). The product is the compound, lead sulfide (PbS). The chemical equation for the reaction is given in Eq. (6).

$$Pb + S \rightarrow PbS \qquad (6)$$

Problem 2. Suppose you have 1.08 gram atoms of lead and 2.81 gram atoms of sulfur. (a) Which of the elements is in excess? How much of that element will be left over when the reaction has gone to completion? (b) How many moles of lead sulfide can be produced?

Solution: First we must obtain the ratios of reactants and products from the chemical equation. Since the coefficients in Eq. (6) are not shown explicitly, we infer that each coefficient is unity:

$$1\ Pb + 1\ S \rightarrow 1\ PbS$$

The desired ratios then follow from the coefficients.

$$\frac{\text{Gram atoms of S that react}}{\text{Gram atoms of Pb that react}} = \frac{1}{1} \quad (7)$$

$$\frac{\text{Moles of PbS that form}}{\text{Gram atoms of Pb that react}} = \frac{1}{1} \quad (8)$$

(a) According to (7), the elements react in a one-to-one ratio of gram atoms. Since you have 1.08 gram atoms of lead and 2.81 gram atoms of sulfur, clearly the sulfur is in excess: the lead will be converted completely to lead sulfide. Let the number of gram atoms of sulfur that *react* be denoted by x. Then, from (7),

$$\frac{x}{1.08} = \frac{1}{1}.$$

Therefore $x = 1.08 \times 1 = 1.08$ gram atoms. Thus, 1.08 gram atoms of sulfur will react. Since the total amount of sulfur is 2.81 gram atoms, $2.81 - 1.08$, or 1.73 gram atoms of sulfur will be left over.

(b) Let the unknown number of moles of lead sulfide be denoted by y. Then, from Eq. (8),

$$\frac{y}{1.08} = \frac{1}{1}.$$

Therefore, $y = 1.08 \times 1 = 1.08$ moles. Thus, 1.08 moles of lead sulfide can be produced.

Weight of Reactants and Products

Although stoichiometric calculations are made most easily in terms of moles, when substances are actually weighed out in the laboratory one uses balances that are calibrated to measure weight in grams. It is not practical to construct a "mole balance" because one mole is a different number of grams for different substances. We must therefore learn how to convert from grams to moles.

The basic equation is Eq. (9), which defines the molecular weight mathematically as the weight per mole.

$$[\text{Molecular Weight}] = \frac{[\text{Weight (in grams)}]}{[\text{Number of moles}]} \tag{9}$$

"The molecular weight equals the weight per mole"

When we know the molecular formula, we can of course calculate the molecular weight from atomic weights, as explained in Chapter 6 (p. 96). Thus, the molecular weight may be treated as a known constant for the given substance. To convert from moles to grams, we rearrange Eq. (9) to the convenient form (10). To convert from grams to moles, we use (11).

To convert from moles to grams:

$$[\text{Weight (in grams)}] = [\text{Molecular Weight}]\,[\text{Number of moles}] \tag{10}$$

To convert from grams to moles:

$$[\text{Number of moles}] = \frac{[\text{Weight (in grams)}]}{[\text{Molecular Weight}]} \tag{11}$$

The conversion from gram atoms to grams, or vice versa, is entirely analogous. The basic equation is (12), which is analogous to (9). Rearrangement of (12) leads to (13) and (14).

$$[\text{Atomic Weight}] = \frac{[\text{Weight (in grams)}]}{[\text{Number of gram atoms}]} \tag{12}$$

To convert from gram atoms to grams:

$$[\text{Weight (in grams)}] = [\text{Atomic Weight}]\,[\text{Number of gram atoms}] \tag{13}$$

To convert from grams to gram atoms:

$$[\text{Number of gram atoms}] = \frac{[\text{Weight (in grams)}]}{[\text{Atomic Weight}]} \tag{14}$$

The conversion from moles to grams, and vice versa, is best illustrated by working some typical problems.

Problem 3. What is the weight (in grams) of 1.08 moles of lead sulfide?

Solution: To convert from moles to grams, we use Eq. (10). The molecular weight of lead sulfide (PbS) is obtained as follows:

$$\underset{\text{At. wt. of Pb}}{207.19} + \underset{\text{At. wt. of S}}{32.06} = \underset{\text{Mol. wt. of PbS}}{239.25}$$

Substituting in (10), we find the weight (in grams) of lead sulfide as follows:

$$\underset{\text{Mol. wt. of PbS}}{[239.25]} \times \underset{\text{No. of moles}}{[1.08]} = \underset{\text{Wt. in grams}}{[258.39\text{ g}]}$$

Problem 4. If we weigh out 100 g of pure sucrose (table sugar), how many moles do we have? The molecular formula of sucrose is $C_{12}H_{22}O_{11}$.

Solution: To convert from grams to moles, we use Eq. (11). The molecular weight of sucrose ($C_{12}H_{22}O_{11}$) is obtained as follows:

$$\underset{12 \times \text{At. wt. of C}}{12 \times 12.01} + \underset{22 \times \text{At. wt. of H}}{22 \times 1.008} + \underset{11 \times \text{At. wt. of O}}{11 \times 16.00} = \underset{\text{Mol. wt. of } C_{12}H_{22}O_{11}}{342.30}$$

Substituting in (11), we find the number of moles of sucrose as follows:

$$\underset{\text{Wt. (in grams)}}{[100]} \div \underset{\text{Mol. wt.}}{[342.30]} = \underset{\text{No. of moles}}{[0.292]}$$

Problem 5. 6.08 g of hydrogen is mixed with 4.11 *moles* of oxygen. The mixture is "sparked" so that reaction takes place and water is produced. (a) Which element is in excess? (b) How many moles of H_2O are produced? (c) How many grams of H_2O are produced?

Solution: (a) *First step: Write the chemical equation and deduce the molar ratios of reactants and products.* This has already been done; see Eqs. (2) to (5).

Second step: Express all reactants in units of moles. The amount of oxygen is already stated in moles. The amount of hydrogen is stated in grams. To convert to moles, we use Eq. (11). The molecular weight of hydrogen (H_2) is calculated as follows:

$$\underbrace{2 \times 1.008}_{2 \times \text{At. wt. of H}} = \underbrace{2.016}_{\text{Mol. wt. of } H_2}$$

Substituting in (11), we find the number of moles as follows:

$$\underbrace{[6.08]}_{\text{Wt. of } H_2} \div \underbrace{[2.016]}_{\text{Mol. wt. of } H_2} = \underbrace{[3.02]}_{\text{Moles of } H_2}$$

Third step: Find which reactant is in excess, and which is the limiting quantity for the reaction. Here we use the molar ratio obtained in (3), which is repeated below.

$$\frac{\text{Moles of } H_2 \text{ that react}}{\text{Moles of } O_2 \text{ that react}} = \frac{2}{1} \tag{3}$$

Thus, 3.02 moles of hydrogen react with half that number of moles, that is, with 3.02/2 or 1.51 moles of oxygen. Since we actually have 4.11 moles, the oxygen is evidently in excess.

(b) *Fourth step: Find the number of moles of product that is formed.* Since the oxygen is in excess, the amount of hydrogen is the limiting quantity for the reaction. To find the number of moles of H_2O that is formed when 3.02 moles of hydrogen react with excess oxygen, we use the molar ratio (4), which is repeated below.

$$\frac{\text{Moles of } H_2O \text{ that form}}{\text{Moles of } H_2 \text{ that react}} = \frac{2}{2} = \frac{1}{1} \tag{4}$$

Since the ratio is simply one-to-one, 3.02 moles of hydrogen will produce 3.02 moles of H_2O.

(c) *Fifth step: Convert moles of product to weight (in grams) of product.* As a final step, we wish to express the weight of water thus produced in units of grams. To convert from moles to grams, we use Eq. (10). The molecular weight of H_2O is 18.016 (see p. 96). Substituting in Eq. (10), we then find the weight in grams as follows:

$$\underset{\text{Mol. wt. of } H_2O}{[18.016]} \times \underset{\text{Moles of } H_2O}{[3.02]} = \underset{\text{Wt. (in grams) of } H_2O}{[54.4 \text{ g}]}$$

Thus, 54.4 g of water are produced.

Problem 5 was rather involved, consisting of five steps. However, none of the steps by itself is uncommonly difficult, and the five-step procedure is of general utility, so that it will pay you to understand the stepwise sequence. In the questions at the end of the Chapter, most of the numerical problems will involve single steps only, but you should realize that a chemist dealing with an actual problem in the laboratory usually has to go through the entire five-step sequence.

The balanced chemical equation

Calculations of the sort described above can give accurate results only if the chemical equation for the reaction is properly balanced, that is, if the equation is written so that there is conservation of atoms for each element. In this section we shall show how to write a balanced equation.

There is a well-known recipe for rabbit stew which begins with the words, "First, catch your rabbit." Similarly, in writing a chemical equation, the first requirement is to know *all* reactants and *all* products —not just qualitatively, but by their precise molecular formulas. Without that information, no balanced chemical equation can be written.

After writing down the molecular formulas for the reactants and products, we proceed to find the molar ratios, that is, the coefficients in the chemical equation. Our only guiding principle here is that there must be conservation of the atoms of each element. The result, of course, is the desired balanced chemical equation. We shall describe two methods for arriving at balanced chemical equations: balancing by inspection, and balancing by the method of algebra. These methods are best explained by working a few problems.

Problem 6. (To illustrate equation-balancing by inspection.) Iodine (I_2) reacts with chlorine (Cl_2) to produce the compound, iodine chloride (ICl). To derive the balanced chemical equation, we begin by writing the molecular formulas in the form of an unbalanced chemical equation:

$$I_2 + Cl_2 \longrightarrow ICl \text{ (not yet balanced)} \tag{15}$$

We then inspect the unbalanced equation, noting all species of atoms that appear. In this case, there are iodine atoms and chlorine atoms. If the equation is to be balanced by inspection, there should be at least one species of atoms that appears in only *one* reactant and *one* product; otherwise balancing by inspection is inadvisable and the method of algebra should be used. In this case, iodine atoms appear in only one reactant, I_2, and one product, ICl. Similarly, chlorine atoms appear in only one reactant, Cl_2, and one product, ICl. Either species of atoms is suitable for starting the balancing process. For definiteness, we shall begin by balancing the iodine atoms.

When balancing by inspection, it is convenient to take one mole of the first reactant we wish to balance. We therefore take one mole of I_2 and rewrite Eq. 15 by inserting the coefficient 1:

$$1\ I_2 + Cl_2 \longrightarrow ICl \text{ (not yet balanced)} \tag{16}$$

To balance the iodine atoms, we must deduce that amount of ICl that is consistent with conservation of iodine atoms: since one mole of I_2 contains two gram atoms of I, we need two gram atoms of I in the product. Since each mole of ICl contains one gram atom of I, we need two moles of ICl. We therefore insert the coefficient 2 before ICl:

$$1\ I_2 + Cl_2 \longrightarrow 2\ ICl \text{ (I-atoms are balanced; Cl-atoms are not balanced)} \tag{17}$$

The resulting Eq. (17) is typical of a partially balanced chemical equation. The molecular formulas of I_2 and ICl are shown with a coefficient, which implies that the equation is balanced with respect to those substances. The molecular formula of Cl_2 is not shown with a coefficient, which implies that the equation is not yet balanced with respect to Cl_2. Note that we use the coefficient as a marker to tell us which molecular formulas have already been balanced. For that reason it is necessary to insert a coefficient whenever a formula has been balanced, even if that coefficient happens to be unity.

As a consequence of balancing the iodine atoms, we are now in a position to balance the chlorine atoms. Two moles of ICl contain two gram atoms of Cl. Thus, to conserve chlorine atoms, we need one mole of Cl_2. We therefore insert the coefficient 1 before Cl_2 in (17) and obtain (18).

$$1\ I_2 + 1\ Cl_2 \rightarrow 2\ ICl \text{ (balanced)} \tag{18}$$

We then see that Eq. (18) is completely balanced because each molecular formula in (18) is shown with a coefficient.

When we balance a chemical equation by inspection, we actually use the methods of algebra, but we are not consciously aware of that fact because we perform the required operations by mental algebra. If the procedure seems complicated, it will be because there are limits to the number of steps of algebra that the average human mind can carry out without the aid of pencil and paper. Unfortunately, many people have a prejudice against the deliberate use of algebra and avoid contact with it like the plague. Such people create unnecessary problems for themselves, because algebra is nothing but a tool that simplifies the solution of certain problems, just as a power-saw simplifies the cutting of a piece of lumber. For our next example, we shall solve Problem 6 by the method of algebra.

Problem 7. Balance the equation for the reaction of iodine (I_2) with chlorine (Cl_2) to produce iodine chloride (ICl) by the method of algebra.

Solution: We begin by writing the molecular formulas for reactants and products in the form of an unbalanced chemical equation.

$I_2 + Cl_2 \rightarrow ICl$ (not yet balanced)

We then, arbitrarily, insert the coefficient 1 before the first molecular formula that appears (in this case, I_2) and insert letters $a, b, \ldots$ before the other molecular formulas—a different letter before each formula.

$$1\ I_2 + a\ Cl_2 \rightarrow b\ ICl \text{ (to be solved for } a \text{ and } b) \tag{19}$$

The letter-coefficients $a, b, \ldots$ are then evaluated by the method of algebra. To do this, we express the conservation of each species of atoms in the form of an equation. In the present case, we write Eqs. (20) and (21).

To conserve iodine atoms in Eq. (19):

$$1 \times 2 = b \times 1 \tag{20}$$

(1 mole of I_2 × 2 g. at. of I per mole) = (b moles of ICl × 1 g. at. of I per mole)

To conserve chlorine atoms in Eq. (19):

$$a \times 2 = b \times 1 \tag{21}$$

(a moles of Cl_2 × 2 g. at. of Cl per mole) = (b moles of ICl × 1 g. at. of Cl per mole)

The two equations, (20) and (21), enable us to solve for a and b.

From (20): $b = 2$ (22)

From (21) and (22): $2a = b = 2$

$a = 1$ (23)

On inserting the results found in (22) and (23) in Eq. (19), we obtain the balanced equation, (24), which of course is identical to Eq. (18) obtained in problem (6) by the method of inspection.

$$1\ I_2 + 1\ Cl_2 \rightarrow 2\ ICl \tag{24}$$

The method of algebra has the advantage that all steps in the balancing process are written down on paper, so that each step can be reexamined and checked at leisure. The method is truly of general applicability: it will work with chemical reactions of all kinds, no matter how complex. The method also enables us to detect possible omissions in the initial list of reactants and products. If that list is incomplete, the algebra will show it, because it will then be impossible to have conservation of atoms for all species of atoms.

Problem 8. As a final example, we wish to balance a somewhat more complicated chemical equation, namely, that for the reaction of phosphorus (P_4) with hydrogen (H_2) to produce the compound, phosphine (PH_3). For a previous discussion of this reaction, see Fig. 6–6.

We begin by writing the unbalanced chemical equation (25).

$$P_4 + H_2 \rightarrow PH_3 \quad \text{(not yet balanced)} \tag{25}$$

Next, we insert the coefficient 1 before P_4 and unknown coefficients *a*, *b* before H_2 and PH_3.

$$1\,P_4 + a\,H_2 \rightarrow b\,PH_3 \tag{26}$$

We then write equations to express the conservation of each species of atoms.

To conserve phosphorus atoms in Eq. (26):

$$1 \times 4 = b \times 1 \tag{27}$$

(1 mole of P_4 × 4 g. at. of P per mole) = (b moles of PH_3 × 1 g. at. of P per mole)

To conserve hydrogen atoms in Eq. (26):

$$a \times 2 = b \times 3 \tag{28}$$

(a moles of H_2 × 2 g. at. of H per mole) = (b moles of PH_3 × 3 g. at. of H per mole)

The two equations, (27) and (28), enable us to solve for *a* and *b*.

From (27): $b = 4$ (29)

From (28) and (29): $2a = 3b = 3 \times 4 = 12$

$$a = 6 \tag{30}$$

On inserting the results in Eq. (26), we obtain the balanced equation, (31).

$$1\ P_4 + 6\ H_2 \rightarrow 4\ PH_3 \tag{31}$$

Note that the coefficients in that equation are in precisely the same ratio as the volumes of the gaseous reactants and products in Fig. 6–6, confirming that the algebra has been correct.

Suggestions for further reading

Brunauer, S., and Copeland, L. E., "Chemistry of Concrete," *Scientific American,* April 1964, p. 80.

Karasek, F. W., "Analytic Instruments in Process Control," *Scientific American,* June 1969, p. 112.

Plumb, R. C., "Chemical Principles Exemplified," *Journal of Chemical Education, 47,* March 1970, p. 175.

Pratt, C. J., "Sulfur," *Scientific American,* May 1970, p. 62.

Stone, J. K., "Oxygen in Steelmaking," *Scientific American,* April 1968, p. 24.

For more detailed treatments of stoichiometry, the reader is referred to more advanced textbooks on general chemistry or analytical chemistry. (See, for example, the textbooks suggested at the end of Ch. 1.)

Questions

1. Benzene is an aromatic liquid with the molecular formula C_6H_6.
(*a*) Find the molecular weight of C_6H_6.
(*b*) Find the number of moles of C_6H_6 in 1.00 g of benzene.
(*c*) Find the weight (in g) of 0.400 moles of C_6H_6.
(*d*) Find the weight of carbon combined in 1.00 g of benzene.

2. Silicon tetrafluoride is a gas with the molecular formula SiF_4.
(*a*) Find the molecular weight of SiF_4.
(*b*) Find the number of moles of SiF_4 in 22.0 g of silicon tetrafluoride.
(*c*) Find the weight (in g) of 1.6 moles of SiF_4.
(*d*) What is the gram molecular volume (in liters, at 0°C and 1 atm) of silicon tetrafluoride? What is the weight of one gram molecular volume of silicon tetrafluoride?
(*e*) What is the density of silicon tetrafluoride gas at 0°C and 1 atm?

3. Glycine, $C_2H_5O_2N$, belongs to a class of substances called *amino acids* that are essential to the human diet.
(*a*) Find the molecular weight of glycine.
(*b*) Find the number of moles in 10.0 g of glycine.
(*c*) Glycine reacts with sodium hydroxide in a one-to-one molar ratio. What weight (in g) of glycine is needed to react with 0.03 moles of sodium hydroxide?
(*d*) A chemist isolates a compound containing 15.2 wt. % of nitrogen. On the basis of that analysis, could the compound be glycine?

4. The reaction of aluminum with oxygen to give aluminum oxide is represented by the following balanced chemical equation:

$$4\ Al + 3\ O_2 \longrightarrow 2\ Al_2O_3$$

(*a*) State three meanings implied by the balanced chemical equation.
(*b*) What is the molar ratio of oxygen to aluminum in that reaction? What is the molar ratio of aluminum oxide to aluminum?
(*c*) Jones, who is a competent chemist, represents the reaction of aluminum with oxygen by the equation:

$$2\ Al + 3/2\ O_2 \longrightarrow Al_2O_3$$

Does the equation written by Jones refer to the same reaction as that written above? Is the equation written by Jones a balanced equation? Is Jones describing the reaction in terms of molecules or in terms of moles?

5. Write balanced chemical equations for the following reactions:
(*a*) Aluminum (Al) plus chlorine (Cl_2) gives aluminum chloride ($AlCl_3$).
(*b*) Magnesium (Mg) plus oxygen (O_2) gives magnesium oxide (MgO).
(*c*) Silver carbonate (Ag_2CO_3) gives silver oxide (Ag_2O) plus carbon dioxide (CO_2).
(*d*) Silver oxide (Ag_2O) gives silver (Ag) plus oxygen (O_2).
(*e*) Sodium chloride (NaCl) in electrolysis gives sodium metal (Na) and chlorine (Cl_2).
(*f*) Methane (CH_4) plus oxygen (O_2) gives carbon dioxide (CO_2) plus water (H_2O).

6. Potassium chlorate ($KClO_3$) decomposes at elevated temperatures to give solid potassium chloride (KCl) and oxygen gas (O_2). How many moles of oxygen are obtained from the decomposition of 6.13 g of potassium chlorate?

7. Calcium carbonate ($CaCO_3$) decomposes at elevated temperatures to give solid calcium oxide (CaO) and carbon dioxide gas (CO_2).
(*a*) Write a balanced chemical equation for that decomposition.
(*b*) How many grams of calcium carbonate must be present if the sample gives off 0.0200 moles of carbon dioxide?

8. The reaction of hydrogen (H_2) with bromine (Br_2) to give hydrogen bromide (HBr) is represented by the following balanced chemical equation:

$$H_2 + Br_2 \longrightarrow 2\,HBr$$

(*a*) How many moles of bromine are required to react with 0.500 moles of hydrogen? How many moles of hydrogen bromide are produced?
(*b*) How many *grams* of bromine are required to react with 0.500 g of hydrogen? How many *grams* of hydrogen bromide are produced?

9. Consider the reaction of magnesium with nitrogen at elevated temperatures, which is represented by the balanced chemical equation:

$$3\,Mg + N_2 \longrightarrow Mg_3N_2$$

For each mole of Mg_3N_2 that is produced,
(*a*) how many gram atoms of magnesium are required?
(*b*) how many grams of magnesium are required?
(*c*) how many gram molecular volumes of nitrogen gas are required?

When you're able to recognize what you're looking at it will not look as it looked when you were unable to.

JAMES GOULD COZZENS (1903–)

Eight

THE STRUCTURE OF ATOMS: 1. THE FAILURE OF THE DALTONIAN ATOM

JOHN DALTON conceived of atoms as indestructible bodies, endowed with a characteristic weight for each element, and capable of combining with other atoms to form molecules. How would such atoms as Dalton's be able to combine to form molecules? What rules might govern the manner of their combination? In this chapter we shall describe some extremely important experiments and observations that will cast some light on these problems. The experiments took place in a relatively short interval of time, at the close of the nineteenth century and the opening of the twentieth, at a time when many thought that the nature of the physical world had already been fully explored. They forced the complete abandonment in scientific usage of Dalton's unstructured atom and produced a drastic revision of our whole concept of the structure of matter.

Static electricity; the concept of electric charge

A complete discussion of the nature of electricity and of the many ways in which it manifests itself is beyond the scope of this book and is not really necessary for the understanding of our central problem. There are, however, a few facts that we must know in order to understand the inadequacies of Dalton's theory.

It has been known since ancient times that certain substances, when rubbed briskly, are capable of attracting and holding tiny bits of paper and other similar materials. Further experiment demonstrates that this attribute is different for different materials. Thus, when hard rubber or amber is rubbed briskly with a piece of fur it will *repel* another piece of rubber or amber that has been treated in a similar fashion, but it will be *attracted* to a piece of glass that has been rubbed with a silk cloth. Similarly, two pieces of glass, when activated in this fashion, will *repel* each other. A large number of experiments of this type have served to demonstrate that two kinds of activation or *charge* are possible. Bodies bearing the same kind of charge repel each other whereas bodies bearing the opposite kinds of charge attract one another. By general agreement, the kind of charge that is produced on a rubber rod by rubbing it with fur is called *negative,* and the type produced on glass by rubbing it with silk is called *positive.* The unit of charge has been defined, and the law that governs the interaction of charged bodies has been worked out. What concerns us at this point is merely that such a phenomenon exists.

Our theory of the nature of matter must include an explanation for the following electrical phenomena: (1) An object, when rubbed against another object made of a different material, generates something that we call *electrical charge;* (2) There are two, and only two, types of charge: positive and negative. The simple kind of experiment that we have described suggests that ordinary matter, which is uncharged, consists of equal amounts of positively charged and negatively charged bodies; and further, that charging by friction, that is, by rubbing together the two objects, is simply the act of *separating* the two kinds of charge. Obviously there is nothing in Dalton's theory to help us explain this.

The behavior of gases carrying an electric current

We are surrounded by examples of another phenomenon which throws a great deal of intellectual light on the ultimate structure of matter. This is the ubiquitous neon sign, a glass tube filled with neon gas at low pressure through which an electric current is passing. That the neon tube also throws light of a visible kind and is used to advertise everything from salvation to corn salve is perhaps one of the more regrettable applications of science.

A simple experiment will illustrate many of the features associated with the passage of electrical currents through gases as the gas pressure is lowered. Consider the apparatus shown in Fig. 8-1. A glass tube about one inch in diameter and perhaps two feet long has metal plates imbedded at the ends and is provided with an outlet through which the gas within the tube can be removed progressively by means of a vacuum pump. The gas may be air, neon, mercury vapor, helium, or any other convenient gas.

Fig. 8-1
Cathode ray tube.

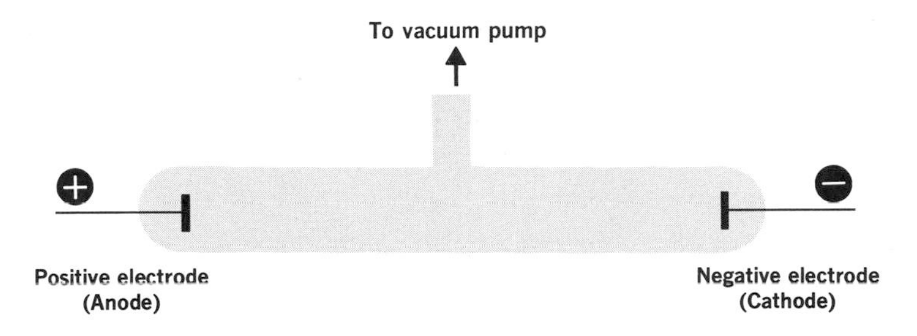

The metal plates connected to a source of high voltage are called *electrodes*. If the electrodes are separated by less than one inch and connected to a high voltage supply, say 15,000 v, the familiar spark will jump the gap. But if the electrodes at high voltages are separated by as much as two feet, as in this experimental tube, and the gas is at atmospheric pressure, no sparking occurs, and no current flows. As the pump progressively removes the gas, however, a series of spectacular events attends the gradual lowering of the pressure. An electric current begins to flow and, at the same time, one sees beautiful streamers of light twisting and turning through the gas between the electrodes. As the pressure becomes lower and lower, the appearance of the tube carrying the electric current changes. It passes through well-defined stages until at last, with the pressure reduced to 1/1,000,000 of an atmosphere, the light disappears completely and there is only a greenish glow emanating from the glass in the region of the positive electrode. If the gas pressure is reduced still further, the current ceases to flow.

We are particularly interested in observations made at pressures around 1/1,000,000 of an atmosphere, where current still flows but only the greenish glow is emitted. The question that naturally arises is:

Sir Joseph John Thomson:
1856–1940
Thomson's researches on the discharge of electricity through gases led to the discovery of the electron and of the existence of isotopes. [Courtesy Brown Brothers.]

How does the current get across the large gap in the electrical circuit? After all, under ordinary circumstances one must have a continuous wire or other electrical conductor in order for electricity to flow.

A series of ingenious experiments, conducted mainly by the British physicist, J. J. Thomson, led to the conclusion that originating near the negative electrode (called the cathode) was a swiftly moving stream of charged particles which moved across the void between the two electrodes. These particles were first called *cathode rays* because of their origin at the cathode, and are now called *electrons.* They have the following properties:

1. The electrons move in a straight line from the cathode (negative electrode) to the positive electrode.
2. They bear considerable energy, since thin metal sheets placed in their path are heated to red heat by the particles hitting them.
3. They each bear a negative charge of electricity, as shown by the fact that they are attracted toward a positively charged object held outside the tube.
4. The mass of the individual electron is extremely small.

It might be suggested that the "electrons" obtained in these experiments are charged *atoms.* But this suggestion must be rejected on the basis of the following evidence: (1) The cathode rays or electrons are the same (have identical properties) no matter what gas is used in the tube, or what metal is used for the electrodes; (2) the mass of the electron is *much* smaller than that of any atom. If the cathode rays were charged atoms, one would expect them to be different for different gases, and one would expect their masses to be consistent with the known weights of the atoms.

J. J. Thomson was able to determine the ratio of the mass to the charge of the electron, and the American physicist, Robert Millikan, in 1909, determined the charge of a single electron, thus making it possible to calculate the mass. It turns out that all electrons have the same mass and charge. In round numbers, the mass of a single electron was found to be only 1/1840 of that of a hydrogen atom. The charge of the electron was the smallest quantity of charge known at that time, and no particle bearing a smaller charge has yet been found.† For this reason, the quantity of charge associated with an electron has been defined as the unit of charge for submicroscopic particles.

Fig. 8-2
Canal rays.

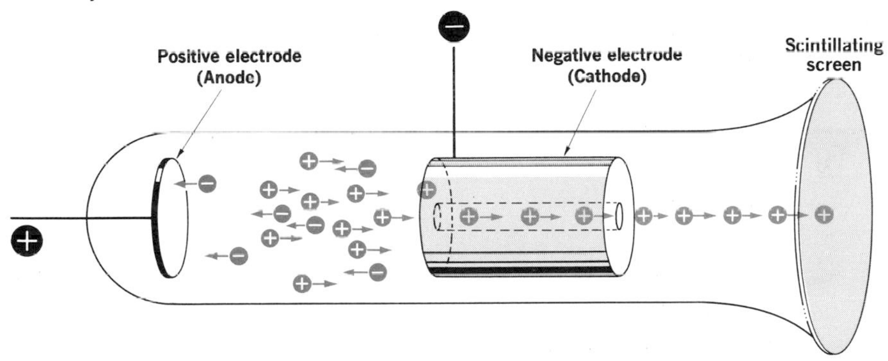

Canal rays

Working with high voltage tubes similar to the one shown in Fig. 8-2, in which a hole was drilled into the cathode, Thomson observed *behind*

† Some theories suggest, however, that particles with a charge smaller than that of the electron, called quarks, may exist.

the cathode another stream of charged particles that had apparently passed through the hole. These particles were, however, quite different from the previously discovered electrons. First of all, they were positively charged, and secondly, they were massive particles, being of the same order of mass as the atoms or molecules of which the gas was composed. Could these particles be what is left of the atom or molecule after an electron (or electrons) has been pulled out of the atom? That this hypothesis is correct is suggested by the fact that the mass of these particles (originally called canal rays, and at present called *ions*†) varies with the gas used in the tube. Thus, if the tube is filled with helium gas at a low pressure, there is produced a stream of electrons moving toward the positively charged electrode and a stream of positively charged helium atoms (helium ions) of atomic weight four, moving in the opposite direction. This process may be symbolized as:

$$\underset{\text{Helium atom}}{He} \xrightarrow[\text{energy}]{\text{Electrical}} \underset{\text{Helium ion}}{He^{\oplus}} + \underset{\text{Electron}}{e^{\ominus}}$$

If the tube contains neon instead, the cathode rays are still the same, but now the canal rays are composed of positively charged neon atoms (neon ions):

$$Ne \xrightarrow[\text{energy}]{\text{Electrical}} Ne^{\oplus} + e^{\ominus}$$

In the case of both helium and neon, the gas molecules are known to consist of just one atom per molecule. When a gas such as nitrogen is used, which ordinarily consists of molecules made up of two atoms, the situation is slightly more complex but still readily understandable in terms of our hypothesis. Several reactions go on simultaneously. They all produce electrons, but now the canal rays consist of two species:

$$N_2 \xrightarrow[\text{energy}]{\text{Electrical}} N_2^{\oplus} + e^{\ominus}$$
$$N_2 \xrightarrow[\text{energy}]{\text{Electrical}} N + N^{\oplus} + e^{\ominus}$$

† Derived from the Greek word for "migratory" to indicate that ions move toward the oppositely charged electrode.

In this case, there are formed positively charged nitrogen molecules (N_2^+), and positively charged nitrogen atoms (N^+), that result from the decomposition of nitrogen molecules. Neutral nitrogen atoms (N) are also formed but are not attracted to either electrode.

It seems an inescapable conclusion that in these experiments a process is taking place that violates Dalton's fundamental postulate that atoms are indestructible. These experiments can be explained only by assuming that under the influence of the strong electrical forces, the atoms of the gas are being broken up into two parts: a tiny negatively charged electron, and a massive positively charged ion.

Isotopes

The discovery of canal rays and of the means for determining their masses produced another revolution in our concept of atoms. Dalton had expressly stated the belief that *all* the atoms of a given element are identical, especially with regard to their mass. However, investigation of the canal rays revealed that, for many elements, atoms are not precisely uniform. Such elements were found to be mixtures of atoms, all alike chemically but differing slightly in atomic mass. For example, neon is a mixture of neon atoms with masses (on the atomic weight scale) of 20, 21, and 22.

Atoms with the same chemical properties but different masses are referred to as *isotopes*. They will be discussed further in Ch. 17.

Natural radioactivity

A third fundamental discovery requiring an alteration of our notions concerning the nature of atoms was made during the same period that saw the discovery of cathode rays and canal rays. This was the discovery in 1895 by Henri Becquerel of the expulsion of various small fragments by the atoms of certain elements. A sample of the element uranium *decays* spontaneously at a slow rate, with the expulsion of fragments called *alpha particles,* each having two units of positive charge and the mass of a helium atom. Other elements decay with the expulsion of *beta particles,* which are identical to the electron or cathode ray discussed in the preceding section. Many of these changes are accompanied by yet a third type of radiation of a very penetrating character which is called *gamma* radiation. It was later found that, in each case, the net result of *radioactive decay* is the formation of a different element. Here we see the atom of Dalton literally crumbling before our eyes.

There is much for us to learn from a detailed study of natural radioactivity, but it must be postponed until we have considered the atom and its structure in more detail. The phenomenon of radioactivity is introduced here only to point up the need for revising Dalton's views concerning the atom, particularly that of its indestructibility. In Ch. 18 we shall return to the problem of radioactive decay.

Rutherford's alpha scattering experiments

One of the immediate results of the discovery of natural radioactivity was that it provided scientists with a new tool for the study of matter, a tool that has since been used with ever-increasing skill and ingenuity. The tiny fragments ejected by atoms undergoing radioactive decay make perfect projectiles for probing the detailed structure of matter. By studying the ways in which alpha and beta particles and gamma rays are deflected, absorbed, or reflected by matter, we can learn much about the nature of the sample under investigation. Further, if these little projectiles are sufficiently energetic, they can dislodge from the atoms small fragments that can be studied, thus enabling us to imagine the structure of the whole atom. This method is analogous to studying the structure of a building by dropping a bomb on it and then studying the pieces. It may seem wasteful, but in the case of atoms it turns out to be one of the most effective methods available.

Very shortly after the discovery of natural radioactivity the three principal radiations emanating from samples of radioactive material were studied intensively, and what was learned about them is summarized conveniently in the following table:

PRINCIPAL RADIATIONS IN NATURAL RADIOACTIVITY

Radiation	Charge	Mass (atomic weight units)	Thickness of layer of air that can be penetrated at one atmosphere
Alpha (α)	2	4	A few centimeters
Beta (β)	-1	1/1840	Hundreds of centimeters
Gamma (γ)	0	0	Several miles

The way in which these radiations behave when they impinge upon very thin sheets of metals (foils) reveals a great deal about the structure of the atoms in the foils. The British physicist, Ernest Rutherford, in 1911, undertook a detailed study of the behavior of highly energetic alpha particles (velocity $= 1.6 \times 10^9$ cm/sec) as they impinged upon a thin gold foil. A schematic drawing of his apparatus is shown in Fig. 8-3.

A deep hole of small diameter is bored in a lead block, and a small sample of a radioactive substance that emits alpha particles is placed in the bottom. This arrangement acts as a gun which "shoots" a continuous stream of projectiles of subatomic size. Immediately in front of the "gun" is placed a series of lead sheets, each pierced with a small hole. These holes are lined up with the hole in the lead block. Thus, only those particles traveling in a particular straight line are allowed to strike the thin sheet of gold, which is the target in this experiment.

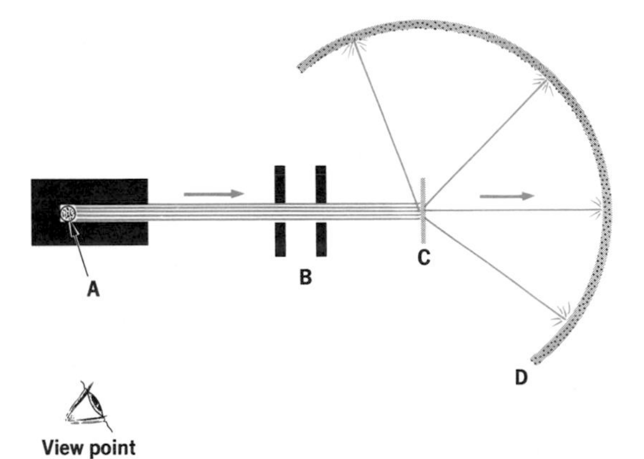

Fig. 8-3
Rutherford's scattering experiment: (a) Radioactive substance emitting alpha particles; (b) lead sheets, pierced with a hole; (c) gold foil, 4×10^{-5} cm thick; (d) scintillating screen, coated with zinc sulfide.

Nothing has been said up to this point about the method of detecting the alpha particles which, being of atomic or subatomic size, are obviously imperceptible to our senses. Many devices have been invented to detect free subatomic particles, the best-known being the Geiger counter. However, the technique employed by Rutherford was more simple and direct; it employed the same principle that is used in luminous watches and clocks. Certain substances, called *phosphors*, emit light when they are bombarded with alpha, beta, or gamma radiation, and each particle as it strikes produces a single flash of light or *scintillation*. In the case of the watch, a phosphor such as zinc sulfide is mixed with a minute amount of radium or some other radioactive substance, and the mixture is used to coat the numerals and hands of the watch. The net effect of many alpha particles striking the phosphor is to produce a steady glow. If, however, the phosphor is coated on a metal plate and alpha particles are shot at it, *each* particle reveals its point of impact by a tiny flash of light that is easily visible in a darkened room with the aid of a low-power microscope.

Ernest Rutherford: 1871–1937 Rutherford is well-known for his theories of radioactive decay and of the structure of the atom. He was the first to transmute one element into another. [Courtesy Culver Service, Inc.]

Thus, by placing a large screen coated with a phosphor behind the gold target (see Fig. 8-3), Rutherford was able to discern the fate of the alpha particles coming from the source. He first looked at the scintillations *without* the target in place and saw that all were located at a single spot. When the target was in place, most of the alpha particles still hit the phosphor at the *same* spot, that is, they were essentially unaffected by the sheet of gold, passing through it as though it were not there. To the person who usually attaches the adjective "solid" to the word "matter," this observation is indeed startling, and it suggests strongly that matter may not be nearly so solid as we had suspected. To Rutherford, however, the behavior of a few of the alpha particles was even more exciting. For these few, instead of passing through the target in a straight line, were deflected from their original paths as they encountered the target. The angles by which the individual alpha particles were deflected varied widely. Some were deflected only by a few degrees, but a very small number actually bounced back in the direction from which they had come (see Fig. 8-3).

On the basis of relatively simple experiments such as this one, an entirely new concept of the nature of the atom could be formulated. These experiments were performed in 1911–1913, and up to that time, no clear model of the atom had emerged. Chemists who thought about the problem usually pictured the atoms as approximately homogeneous solid spheres which somchow arc ablc to form bonds to other atoms. When electrons and canal rays were discovered, that picture was modified just enough to account for the new observations. J. J. Thomson suggested that the atom consists of a sphere of positively charged matter in which negatively charged electrons are embedded, with the number of electrons just sufficient to neutralize the positive charge. His model was rather like a pudding in which are stuck a number of raisins.

Rutherford's experiment makes clear that a "pudding" model of atomic structure just won't work. We already must assume that the atoms are almost touching one another in the solid state. How then could they be penetrated by a particle of comparable atomic weight, if each atom were a solid sphere? And if, somehow, this penetration could be explained, why would some particles fail to penetrate, and others be deflected sharply from their original paths?

Rutherford's model of the atom

Rutherford suggested that the observed *scattering* of the alpha particles by the solid matter of the gold foil can be explained easily if one assumes that on the submicroscopic level matter is not solid at all. He pictured the atom as consisting of a dense central *core* or *nucleus* in which all of the positive charge and nearly all of the mass of the atom are concentrated. Surrounding the nucleus, he imagined the electrons as moving in regularly defined orbits, just as the planets move in regular orbits around the sun (see Fig. 8-4). The orbit of the outermost electron then defines the geographical extent of the atom, just as the orbit of Pluto defines the geographical extent of the solar system. The diameter of the nucleus is very much smaller than that of the entire atom, just as

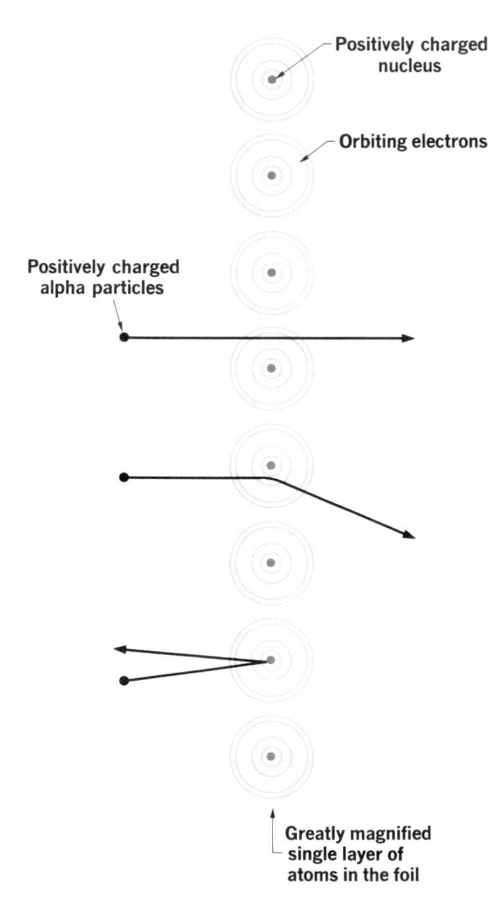

Fig. 8-4
Rutherford's interpretation of the alpha scattering experiment (not to scale). The alpha particles are deflected only if they come close to a nucleus.

the diameter of the sun is much smaller than that of the solar system. It has been found that the diameter of the nucleus is actually about 1/10,000 that of the entire atom.

If one assumes further that an alpha particle is comparable in size to an atomic nucleus, then it becomes clear why most of the alpha particles will have little trouble in passing straight through the thin foil of solid matter. There is only a small chance that the alpha particle, passing through the foil, will collide with an atomic nucleus, or come close enough to a nucleus to be deflected from its original path. Most of the alpha particles will pass through the space between the nuclei where, at worst, they might collide with an electron. However, collision with an electron will have little effect on the motion, because the fast-moving alpha particle weighs nearly 8000 times as much as the electron and hardly "feels" the collision at all. The effect is analogous to that of a 2000 lb. automobile colliding with a baseball.

Rutherford assumed in his calculations that the large deflections were due to the electrical repulsion between the positively charged alpha particle and the atomic nucleus, which he assumed also to be positively charged. He could account for the magnitude of the observed deflections only if the positive charge of the nucleus is concentrated in a very small volume, and if the charge is of a characteristic magnitude for each element. His experiments indicated that all atoms of a given element have the same nuclear charge, and that this charge is different for each element. But the experimental accuracy necessary for a *precise* determination was not achieved until 1920, by the English physicist, James Chadwick. By then several other methods for measuring nuclear charge had been devised. All methods led to the same conclusions:

1. The atoms of the elements must be thought of as consisting of a positively charged nucleus in which the mass is concentrated, and a number of negatively charged electrons orbiting around the nucleus.
2. In the neutral atom, the number of electrons is just sufficient to balance the positive charge of the nucleus.
3. The charge of the nucleus (and hence the number of electrons in the neutral atom) is different for each element.

It was found experimentally that the charge of the nucleus *usually* increases with the atomic weight of the element. The only exceptions to this rule are the pairs Co and Ni, Te and I, and Th and Pa, for which the nuclear charge increases while the atomic weight decreases.

Atomic numbers

Let us again define the unit of negative charge as that possessed by one electron. The unit of positive charge then is an amount of charge equal in magnitude but opposite in sign. Using that definition, the description of the atoms of the elements becomes quite simple. It is found experimentally that the charge on the nucleus of a hydrogen atom is plus one (+1), that on the nucleus of a helium atom is plus two (+2), on a lithium atom plus three (+3), and so on. Thus, the nuclear charge of the elements increases by units of one as one goes from a given element to the next heavier one. (The three exceptions to this rule are noted above.) Because the nuclear charge of an element is always a simple integer, it is usually referred to as the *atomic number* of that element. Today there are 105 known elements with atomic numbers from 1 to 105.

A few examples of atoms as pictured by Rutherford will help to make these concepts more concrete. Consider first the simplest atom, that of the element hydrogen. According to Rutherford, a hydrogen atom consists of a nucleus with a mass of 1 and charge of +1, with a single electron moving around the nucleus at an average distance such that the average diameter of the diatomic hydrogen molecule is that found in Ch. 5, namely, 2.2×10^{-8} cm. Similarly, an atom of helium consists of a nucleus with mass 4 and charge +2, with *two* electrons orbiting at such distances as to give the atom a diameter of 1.6×10^{-8} cm., which is consistent with the known properties of helium gas. In general, an element with the atomic number Z will have atoms with a charge of +Z on the nucleus, and with Z electrons traveling about the nucleus in stable orbits.

According to Dalton's theory, all atoms of the same element have the same atomic weight, which is characteristic of that element alone. That theory is no longer tenable, however, owing to the discovery of isotopes. Fortunately, Rutherford's model provides us with a new atomic property that is characteristically different for each element, namely, the nuclear charge or atomic number. If we accept the notion that the chemical behavior of an element is determined by the nuclear charge of its atoms rather than by the atomic weight, then atoms will be chemically identical as long as they have the same nuclear charge, even though they might differ in weight.

Our model of the submicroscopic atom is now much more complicated than Dalton ever imagined, and it goes a long way toward explaining the macroscopic behavior of the elements and their compounds. For example, the presence of positive and negative charges within the atom suggests an explanation for static electricity, and also a possible mechanism for the binding of atoms into molecules. It also suggests that matter will have properties we have not yet considered, but which are capable of experimental investigation. Thus, we might predict that atoms can *gain* electrons to form negatively charged ions. Such predictive power marks it as a useful model or theory. However, the Rutherford model as presented here still leaves us with a number of problems. For example, nothing has been said about why the electron is not attracted to the surface of the nucleus, or why the atoms do not penetrate each other's boundaries if they are truly open in structure. These and related problems will be considered in the next chapters.

Suggestions for further reading

Birks, J. B. ed., *Rutherford at Manchester* (New York: W. A. Benjamin Co., Inc., 1963).

da Costa Andrade, E. N., "The Birth of the Nuclear Atom," *Scientific American,* Nov. 1956, p. 93.

Duncan, R. K., *The New Knowledge* (London: Hodder and Stoughton Limited, 1909).

Eve, A. S., *Rutherford* (New York: The Macmillan Company, 1939).

Steinherz, H. A., and Redhead, P. A., "Ultrahigh Vacuum," *Scientific American,* March 1962, p. 78.

Thomson, J. J., *Recollections and Reflections* (New York: The Macmillan Company, 1937).

Questions

In questions 1–8, complete the statement by supplying the missing words or phrases.

1. Electrically charged atoms are called ________________ .

2. The two electrodes in a cathode ray tube are called ________________ and ________________ .

3. The cathode bears an electric charge of ________________ sign and this attracts particles bearing a ________________ electric charge.

4. Negatively charged electrons are attracted to the ________________ .

5. The path of a fast moving charged particle is bent by ________________ .

6. Atoms of different mass, derived from the same element, are called ____________ .

7. The expulsion of small fragments by the atoms of certain elements is called ________________ .

8. Alpha particles have a charge of ________________ and a mass of ________________ . Beta particles have a charge of ________________ and a mass of ________________ , which are properties identical with those of ________________ . Gamma particles have a charge of ________________ and a mass of ________________ . These particles can penetrate layers of air of approximately the following thickness:
Alpha, ________________; Beta, ________________; Gamma, ________________ .

9. What is the significance of the atomic number of an element?

In questions 10–13, use the following atomic numbers: hydrogen, 1; lithium, 3; carbon, 6; nitrogent, 7; sodium, 11; uranium, 92.

10. State the nuclear charge and the number of electrons for the following atoms: hydrogen, sodium, uranium.

11. State the nuclear charge and the number of electrons for (a) the nitrogen atom, and (b) the positively charged nitrogen ion, N^+.

12. Draw diagrams of the following atoms, using Rutherford's model:
(*a*) hydrogen
(*b*) lithium
(*c*) carbon
(*d*) sodium

13. (*a*) What is the total weight (in grams) of all the electrons in one gram atom of uranium? (*b*) Using the atomic weight, 238.03, for uranium, what is the weight percent of the electrons in pure uranium?

14. Do the experiments described in this chapter prove that Dalton's theory is completely wrong? Which parts of Dalton's theory remain intact?

15. What does Rutherford's alpha-scattering experiment tell us about the size of alpha particles as compared to the size of atoms?

16. How do we know that the electrons observed by Thomson are not simply negatively charged atoms?

17. How is the production of static charges of electricity explained in terms of the Rutherford model of the atom?

18. (*a*) Explain how you might determine the *sign* of an unknown charge of static electricity. (*b*) Describe an experiment designed to determine the sign of the charge on an electron.

If you have had your attention directed to the novelties in thought in your own lifetime, you will have observed that almost all really new ideas have a certain aspect of foolishness when they are first produced.

ALFRED NORTH WHITEHEAD (1861–1947)

Nine

THE STRUCTURE OF ATOMS: II. BOHR'S THEORY OF THE HYDROGEN ATOM

WE HAVE SEEN that Rutherford's model of the atom was based on a series of ingenious and decisive experiments that seemed to rule out all other alternatives. Yet this model raised some very serious theoretical problems; according to the laws of macroscopic physics of the time, such an atom should not have been stable. In this chapter, we shall discuss these problems and show how the Danish physicist, Niels Bohr, proposed to resolve them for the hydrogen atom.

Today, Bohr's theory of the hydrogen atom is in many ways obsolete. Nevertheless, the story of the origin and nature of Bohr's theory is worth telling, because it provides an almost perfect example of the interplay of theory and experiment—of the rise, modification, and eventual decline of a useful theory in the face of new experimental facts. To understand this sequence of events is to understand the "scientific attitude."

Bohr's approach was based on thought-provoking ideas about the behavior of matter on the submicroscopic level, now called the *quantum theory,* that had been proposed about a decade earlier by the German physicist, Max Planck (1858–1947). Planck's ideas had been slow in gaining acceptance among physicists because they envisaged a behavior for matter on the submicroscopic level that in many ways differed radically from the observable behavior of such familiar objects as billiard balls and pendulums.

The problem of electromagnetic radiation

What, then are the problems that were raised by Rutherford's model of the atom? Rutherford's experiments, during 1911–1913, indicated that an atom is composed of a heavy central nucleus bearing a positive charge, which is surrounded by negatively charged electrons of much smaller mass. These electrons exist at some distance from the nucleus and from each other, so that most of the space inside the atom is empty. The experiments could not give an answer to the question of whether the electrons were moving or standing still. Yet it was clear to Rutherford on the basis of other considerations that the electrons were necessarily in motion. If, at a given moment, an electron were at rest, the force of attraction that acts between opposite charges would immediately accelerate the electron toward the nucleus and cause it to collide with the nucleus.

Rutherford assumed that the electrons travel in an orbiting motion around the nucleus—a submicroscopic analog of planetary motion around the sun. This means that the electric charges (the electrons) are moving under the influence of a central force (the nucleus); under such circumstances electrodynamic principles should apply. Those principles had been expressed in elegant and very general form by the British physicist, James Clerk Maxwell (1831–1879). Maxwell had shown that electricity, magnetism, and light can all be treated in terms of a single theory. His *electromagnetic theory* (published in the 1860's) summed up everything that was then known concerning light, electricity, and magnetism, and it culminated in a set of equations, now known as *Maxwell's equations,* that expressed quantitatively the relationships among these phenomena. The theory had successfully guided the German physicist, Heinrich Hertz, in 1889, to produce radio waves for the first time by setting electric charges into periodic motion. And the theory predicted quite unequivocally that electric charges moving under the influence of a central force should lose energy by emitting electromagnetic radiation.

Thus, according to electromagnetic theory, an atom consisting of electrons orbiting around the nucleus should continually emit electromagnetic radiation of the same frequency as that of the orbital motion, thereby losing energy. As the atom loses energy, the electrons should move in ever smaller orbits, spiralling inward toward the nucleus and, sooner or later, cause the atom to collapse. In other words, if Maxwell's theory applies to atoms, then the Rutherford atom cannot be stable. (The motion of an electron would be similar to that of a man-made satellite whose orbit passes through the earth's upper atmosphere. The satellite loses energy by friction with the atmosphere and spirals toward the earth's surface.)

It is difficult to overstate the dilemma facing Rutherford and his contemporary physicists. It is easy enough for the layman to say: "We know that atoms are stable. Experiment indicates that atoms consist of a nucleus with electrons orbiting around it. Maxwell's equations predict that atoms of such a structure cannot be stable. Ergo: Maxwell's equations do not apply to atoms." But a professional physicist cannot do this lightly. Physicists are justly proud of the great generality of their science. Cosmologists assume (indeed, they *must* assume) that the laws of terrestrial physics apply throughout the universe. To admit that the same laws do not apply to the submicroscopic universe is a serious matter indeed. Yet at the time of Rutherford's experiments, the first steps had already been taken to develop a viewpoint and construct a theory that would apply specifically to the submicroscopic world.

Some properties of light

Before beginning the discussion of Bohr's solution to this dilemma, let us digress briefly to review some of the general properties of light.

Light is the visible component of electromagnetic radiation. It was discovered by Newton that, when a narrow beam of white light is passed through a prism, the rays of various colors of which the beam is composed are bent at different angles; and the light emerges in the form of a band of colors shading from red at one end to violet at the other. This process of breaking up a light beam is called *dispersion,* and the display of component colors that results is called a *spectrum.* Dispersion can also be produced by falling raindrops, in which case the spectrum is the familiar rainbow.

In physical discussions of the spectrum, it is usual to discard the concept of *color,* which is subjective, in favor of the concept of *wave-*

length, which can be measured objectively. It has been found that light on the macroscopic level can be treated as a wave phenomenon. Thus, Fig. 9–1 shows the wave associated with a beam of light of uniform wavelength. The solid line shows the wave at a given instant. As the beam moves to the right with the speed of light, the wave also moves to the right with the speed of light, and the white line shows the same wave at a later instant.

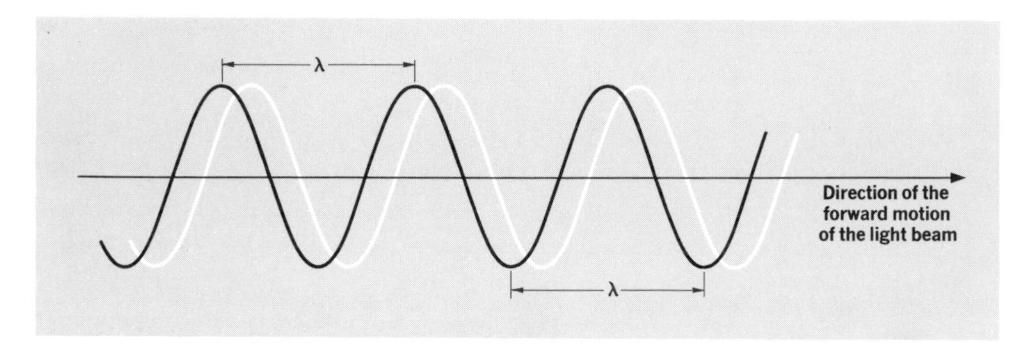

Fig. 9-1
A light wave.

The wavelength, λ (lambda), is the distance between adjacent peaks, troughs, or any other pair of equivalent points in adjacent portions of the wave at a given instant. The frequency, ν (nu), is the number of peaks (or troughs) passing a given point during one second as a result of the forward motion of the wave. Reflection will show that wavelength and frequency are related to the speed of light, c, as in Eq. (1).

$$\lambda\,(\text{cm}) \cdot \nu\left(\frac{1}{\text{sec}}\right) = c\left(\frac{\text{cm}}{\text{sec}}\right) \tag{1}$$

Highly accurate physical instruments are available for measuring the wavelength of light and for producing beams of light of known wavelength. In experimental investigations involving light, it is therefore more usual to specify the wavelength than the frequency. Visible light varies in wavelength approximately from 4×10^{-5} cm (perceived as violet by most people) to 8×10^{-5} cm (perceived as red by most people). If the wavelength is less than 4×10^{-5} cm, the radiation is no longer visible and is called ultraviolet light. If the wavelength is greater than 8×10^{-5} cm, the radiation is in the infrared region of the spectrum and is sensed as heat. Other kinds of electromagnetic radiation are radio waves, whose wavelength is on the order of several meters, and X-rays, whose wavelength is comparable to the size of atoms.

When a beam of light traverses different media, the wavelength and speed change with the medium, but the frequency remains constant. The speed of light is greatest in a perfect vacuum, but it decreases only very slightly when the light beam travels in air or in any other transparent gas. The speed of light in a vacuum is identical for all wavelengths and is one of the fundamental constants of physics. It has been measured repeatedly and with great accuracy. The currently accepted "best value" is 2.997925×10^{10} cm/sec, or very close to 3×10^{10} cm/sec.

Planck's quantum hypothesis

Even before the problem of the hydrogen atom became evident, it was clear that there was something peculiar about the interaction of electromagnetic radiation with matter. Every now and then, a phenomenon would be discovered that could not be readily explained in terms of the accepted physical concepts of the time. Ordinarily, this would not be alarming. Nature is so complicated, and the mathematical tools of physics are frequently so difficult to apply, that it is not always obvious just how to go about explaining a phenomenon. It may take years to find the right approach. But once that approach has been found, the solution to the problem becomes obvious.

What was different about the problem of the interaction of radiation with matter was that these phenomena had occupied the attention of some of the best brains of the time for a good many years, yet no real solution was in sight. Finally, in 1900, the German physicist, Max Planck, decided that the fault lay not with physicists but with physics itself, that is, the accepted concepts of physics were somehow inadequate. To interpret some puzzling facts concerning radiation emitted by hot bodies, he introduced a radically new hypothesis, the *quantum theory*, which presented quite an unconventional picture of matter at the submicroscopic level. Planck was interested in the general phenome-

Max Planck: 1858–1947
[Photograph courtesy Brown Brothers.]

non of incandescence: as the temperature of a solid is raised, the solid emits not only heat but also visible light, first of a dull-red quality, then orange in color, then bright yellow, and finally, at very high temperatures, an intense bluish-white. These color changes could not be reconciled with concepts of classical physics.

Planck assumed that in any substance capable of emitting or absorbing light, there are submicroscopic oscillators which are responsible for this emission or absorption. If the oscillator gains energy, light is absorbed; if the oscillator loses energy, light is emitted. So far he was on familiar ground. But then he broke with traditional physics by assuming that *only certain amounts* of energy can be absorbed or emitted. Thus, if any oscillator oscillates at the frequency ν, the permitted values of the amount of energy that can be absorbed or emitted are given by equation (2).

$$\text{Permitted values of the energy} = nh\nu \qquad (n = 0, 1, 2, \ldots) \tag{2}$$

Equation (2) bears closer examination. The integer n is either zero, or one, or two, or any other positive integer, and h is the constant of proportionality that Planck was able to evaluate on the basis of experimental data. Equation (2) therefore generates a set of discrete energy values that differ by a fixed amount, $h\nu$, which is proportional to the oscillator frequency. By applying Eq. (2) Planck was able to give a precise description of the experimental facts concerning incandescence.

It is convenient to represent the permitted values predicted by Eq. (2) in an *energy level diagram*, such as that shown in Fig. 9-2. In such a diagram, the energy is plotted along the y-axis, and each permitted value of the energy is represented as a horizontal line, drawn at the appropriate energy level. The energy of the lowest level is taken as zero. The levels above the lowest are referred to as excited levels. The permitted energy states are called *quantum states.* The number, **n**, in the expression $nh\nu$ for the energy of the level is called the *quantum number.*

According to Planck, an oscillator whose energy is different from one of the permitted values is never observed. If there were such an oscillator, it could not exist for more than a fleeting moment. It would immediately lose energy and proceed to one of the permitted energy states.

In modern theories of the submicroscopic world, Planck's idea that only certain values of the energy are permitted is generally accepted. Planck's constant, h, appears often in the equations and has become one of the fundamental constants of physics. If we express energy in calories and frequency in cycles per second (cycles/sec), the value of h is 1.583×10^{-34}.

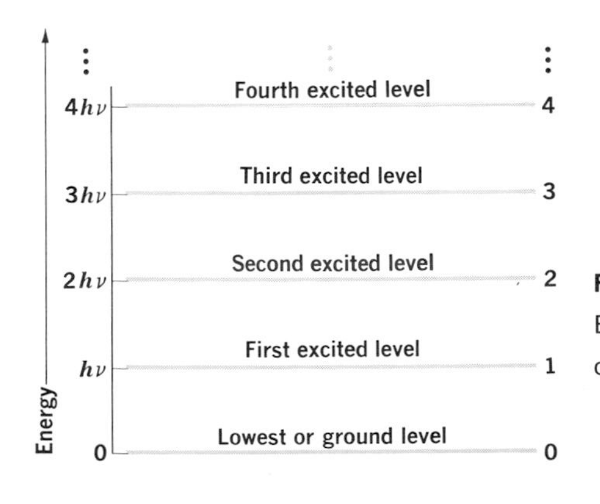

Fig. 9-2
Energy level diagram for a Planck oscillator.

Einstein's theory of the photoelectric effect

Another puzzling phenomenon whose explanation required a break with traditional physics is the *photoelectric effect.* Light of the appropriate wavelength, on striking a metal surface, is capable of ejecting electrons from that surface. It was found early that the values of the energy of the ejected electrons do not depend at all on the *intensity* of the light but *only* upon its wavelength. A more intense light merely produces more electrons: it is necessary to reduce the wavelength in order to increase the energy.

Albert Einstein drew some important inferences concerning the nature of electromagnetic radiation from these observations. He believed that a beam of light could be imagined on the submicroscopic level as a stream of many tiny elementary quantities, called *photons.* That is to say, a photon is related to the macroscopic phenomenon, light, much as an electron is related to the macroscopic phenomenon, electric charge, or as a molecule is related to the macroscopic phenomenon, matter. And just as atoms and molecules gain or lose charge in units of one electron at a time, so Einstein believed that atoms and molecules absorb or emit radiation in units of one photon at a time. He then argued that the energy of a photon must be equal to $h\nu$, where h is Planck's constant (the same constant as in Eq. 2), and ν is the frequency of the macroscopic radiation associated with the photon.

$$\text{Energy of photon} = h\nu \qquad (3)$$

Since the frequency varies inversely with the wavelength (see Eq. 1), light of a shorter wavelength consists of the more energetic photons.

To explain the photoelectric effect, Einstein postulated that each ejected electron is produced by the absorption of one photon, and that in this process the energy of the photon is imparted to the electron. The electron thus gains sufficient energy to leave the metal.

Bohr's theory of the hydrogen atom

It was a major breakthrough for chemistry when Niels Bohr demonstrated (in 1913, when he was only twenty-eight years old) that Planck's

Niels Bohr: 1885–1962
[Photograph courtesy Brown Brothers.]

quantum hypothesis and Einstein's conception of photons can be fruitfully applied to atoms.

Bohr, at the time, was searching for a theory that would reconcile Rutherford's atomic model with the undoubted fact that atoms exist in stable states without emitting radiation. He was especially interested in the hydrogen atom because it is the simplest atom of all. He realized that an electron circling repeatedly around a nucleus would be analogous to the electrical oscillator that Planck had envisioned to account for the behavior of incandescent solids. In particular, the frequency with which the electron completes its circular path would be analogous to Planck's oscillator frequency, ν. (See Eq. (2).) Thus it seemed natural to him to carry the analogy further. He proposed that there is a set of "permitted" energy states in which the atom can exist without emitting radiation. And when he assumed that in these states the energy of the atom is related to the frequency of the electron's motion according to

Eq. (4), he was able to give a remarkably accurate description of many known facts concerning the hydrogen atom.

Permitted values for the energy of the hydrogen atom:

$$E = nh \cdot \frac{\text{Frequency of circular motion}}{2} \tag{4}$$

$$(n = 1, 2, 3, \ldots)$$

Equation (4) bears an obvious and striking resemblance to the basic equation in Planck's quantum theory, Eq. (2), and was, in fact, inspired by the latter. Let us compare the two equations. Both equations state that the energy can have only certain discrete values which are proportional to an integer (the quantum number n) and a frequency; and both equations involve exactly the same constant, h. Bohr's equation also contains an additional factor, $\frac{1}{2}$, but this point of difference is not fundamental.

It can be shown (by application of Newton's laws of motion) that the frequency of motion of an electron of charge e and mass m, moving in a circular orbit of radius r, is $e/(2\pi\sqrt{mr^3})$. Then, if we consider the energy E of the hydrogen atom to be the sum of kinetic and potential energy, this frequency can be related to the energy (Eq. (5)).

$$\text{Frequency of circular motion} = \frac{1}{\pi e^2}\sqrt{\frac{-2E^3}{m}} \tag{5}$$

Equations (4) and (5) can be combined to give an expression for the energy of the hydrogen atom in terms of the quantum number n.

$$E = -\frac{2\pi^2 me^4}{h^2 n^2}; \qquad n = 1,2,3, \ldots \tag{6}$$

Bohr's interpretation of Eq. (6) was that each value of n corresponds to one of the "permitted" energy states of the hydrogen atom. We shall see in the next section that Bohr was able to predict the energy of these states with great accuracy. But first let us consider some other properties of the hydrogen atom, as predicted by Bohr. The equations derived by him are shown in the following table.

PROPERTIES OF THE HYDROGEN ATOM ACCORDING TO BOHR'S THEORY

Energy	$E = -\dfrac{2\pi^2 me^4}{h^2 n^2} = -\dfrac{5.208 \times 10^{-19}}{n^2}$ cal/atom
	$= -\dfrac{313{,}700}{n^2}$ cal/g-atom
Frequency	$f = \dfrac{4\pi^2 me^4}{h^3 n^3} = \dfrac{6.58 \times 10^{15}}{n^3}$ cycles/sec
Orbital Radius	$r = \dfrac{h^2 n^2}{4\pi^2 me^2} = 0.5292 \times 10^{-8}\, n^2$ cm
Velocity	$v = \dfrac{2\pi e^2}{nh} = \dfrac{2.188 \times 10^8}{n}$ cm/sec

From these general formulas one can calculate the properties of specific quantum states. The results obtained for the first few quantum states are shown below.

PROPERTIES OF SPECIFIC QUANTUM STATES OF THE HYDROGEN ATOM

Orbit	Quantum number	Radius (cm)	Velocity (cm/sec)	Frequency (cycles/sec)	Energy (cal/g-atom)
First	1	0.529×10^{-8}	2.188×10^8	6.58×10^{15}	$-313{,}700$
Second	2	2.116×10^{-8}	1.094×10^8	0.82×10^{15}	$-78{,}400$
Third	3	4.76×10^{-8}	0.729×10^8	0.244×10^{15}	$-34{,}900$
Electron and nucleus very far apart	∞	∞	0	0	0

We can see from the table that the first state, with quantum number 1, has the lowest (most negative) value of the energy and is, therefore, the ground state. The energy gap between the ground state and any excited state is very large, so that the ground state is by far the most stable state in the set. The velocity of the electron in the ground state of the hydrogen atom is predicted to be 2.188×10^8 cm/sec. This enormously high value is about a thousand times greater than the average velocity of the hydrogen *molecule* at room temperature (p. 80), and nearly one-hundredth of the velocity of light. The frequency of the electron's motion is predicted to have the fantastically high value of 6.58×10^{15} cycles/sec. This frequency is large even when compared to the frequency of collisions experienced by a hydrogen atom, which we estimate as about 10^{10} per second at a pressure of one atmosphere in the gas phase. Therefore, an electron can traverse its orbit many times between atomic collisions.

Of equal interest is the radius of the electron's orbit, which should be a fair index of the radius of the hydrogen atom. Here we have the opportunity to check Bohr's theory, because we can obtain an independent estimate of the size of the hydrogen *molecule* on the basis of kinetic molecular theory. The average diameter of the *diatomic* hydrogen molecule was thus listed on p. 80 as 2.2×10^{-8} cm. By comparison, the diameter of Bohr's first orbit (twice the radius) is 1.06×10^{-8} cm, or about half of 2.2×10^{-8} cm. We conclude that Bohr's theory certainly leads to plausible values for the atomic dimensions.

The spectrum of the hydrogen atom

The most convincing evidence for the soundness of Bohr's theory comes from the spectrum of the hydrogen atom. When a transparent tube is filled with hydrogen gas at about 0.001 atmosphere pressure and electrical energy is supplied, the hydrogen gas emits a rose-colored glow. When a beam of this light is dispersed by a prism, a spectrum consisting of a number of sharp lines is obtained. A photograph of this spectrum is reproduced in Fig. 9-3. We can see that the light consists of certain discrete wavelengths only. The rest of the spectrum is dark. A spectrum of this sort is known as a *line spectrum* and is, of course, strikingly different from the continuous spectrum or rainbow obtained from the sun or other incandescent bodies.

There is good evidence that the characteristic line spectrum emitted under these conditions is entirely due to hydrogen *atoms.* The input of electrical energy provides a steady stream of electrons moving with very high kinetic energies inside the tube. When such an electron collides with the diatomic hydrogen molecule, enough energy is transferred to cause the molecule to break apart. Thus, there is a continuous supply of energy-rich hydrogen atoms.

Bohr's theory gives a natural explanation for the fact that the hydrogen spectrum consists of discrete lines, and it predicts the wave-

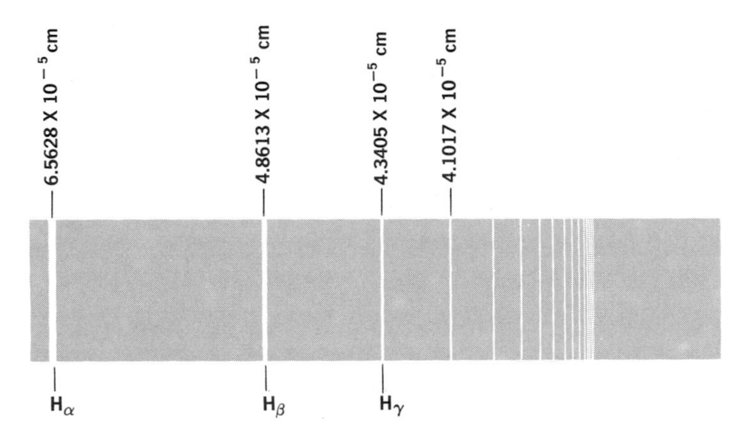

Fig. 9-3
Visible portion of the line spectrum of the hydrogen atom. [From G. Herzberg, *Annalen der Physik* (4) *84* (1927), 565.]

lengths of these lines with remarkable accuracy. To show this, let us begin by drawing a diagram of the permitted energy levels of the hydrogen atom as predicted by Bohr. This is done in Fig. 9-4, in which levels have been calculated from Eq. (6). According to the quantum theory, the hydrogen atom can exist in any one of these permitted energy states without emitting radiation. However, when the atom "drops" from an upper level to a lower one, a photon of radiation of the requisite energy is emitted. (See Eq. (3).) Conservation of energy requires that the energy of the photon be equal to the energy lost by the hydrogen atom; hence we obtain Eq. (7).

$$E_{\substack{\text{Upper}\\\text{state}}} - E_{\substack{\text{Lower}\\\text{state}}} = h\nu = \frac{hc}{\lambda} \tag{7}$$

To derive the entire line spectrum of hydrogen, Bohr assumed that the energy-rich hydrogen atoms are distributed over all possible quantum states. When a hydrogen atom in an excited state emits radiation, the final state can be any state of lower quantum number. For example, an atom with $n = 4$ can drop either to $n = 3$, $n = 2$, or $n = 1$ with the emission of a photon whose energy is characteristic of the process. Therefore, the spectrum should consist of a number of discrete lines,

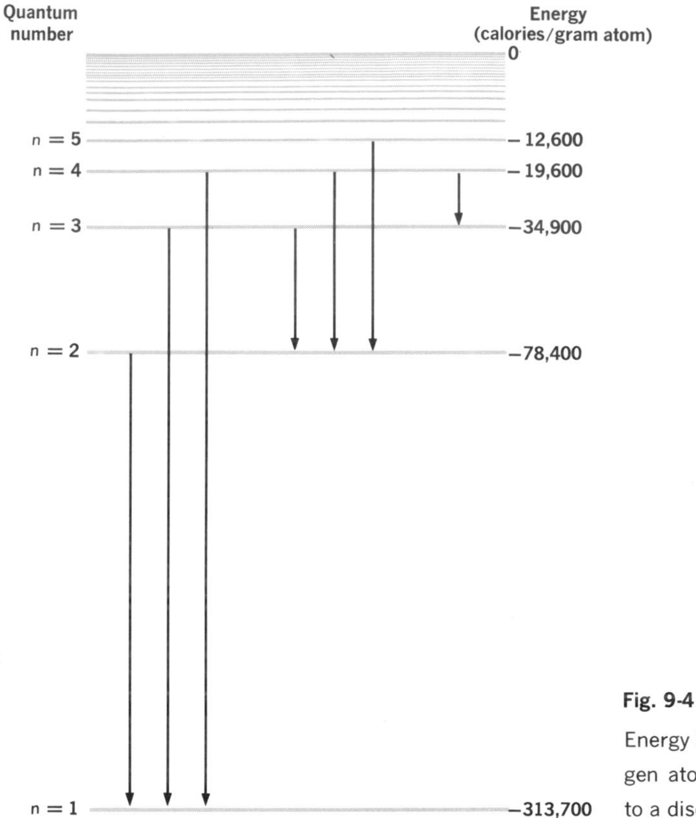

Fig. 9-4

Energy level diagram of the hydrogen atom. Each line corresponds to a discrete line in the spectrum.

one line for each conceivable combination of an upper state with a lower state. Some of these combinations are indicated by arrows in Fig. 9-4.

On this basis, it is a simple matter to calculate the wavelengths of the spectral lines. For definiteness, consider that the upper state has quantum number 2 and energy E_2, and that the lower state has quantum number 1 and energy E_1. (See Fig. 9-5.)

To calculate the wavelength of the corresponding spectral line, Eq. (7) is rearranged to (8):

$$\lambda = \frac{hc}{E_{\substack{\text{Upper} \\ \text{state}}} - E_{\substack{\text{Lower} \\ \text{state}}}} \tag{8}$$

We wish to express E in cal/g-atom and λ in cm; hc must then be taken as 2.859 if the spectrum is measured in air or in an evacuated apparatus.

Referring to the table on p. 151, we find that $E_2 = -78{,}400$ and $E_1 = -313{,}700$; hence, $E_2 - E_1 = -78{,}400 - (-313{,}700)$, or $+235{,}300$ cal/g-atom. On applying Eq. (8), we find that $\lambda = 2.859/$-235,300 or 1.215×10^{-5} cm. Light of that wavelength is actually found in the ultraviolet region.

Similar calculations have been made for some other combinations of quantum states, with the results given below.

SPECTRUM OF THE HYDROGEN ATOM

Quantum numbers (upper state → lower state)	Wavelength predicted by Bohr's theory (cm)	Actual wavelength of nearest spectral line (cm)	Spectral region
→ 1	1.215×10^{-5}	1.2157×10^{-5}	Ultraviolet
3 → 1	1.025×10^{-5}	1.0257×10^{-5}	Ultraviolet
4 → 1	0.972×10^{-5}	0.9725×10^{-5}	Ultraviolet
3 → 2	6.562×10^{-5}	6.5628×10^{-5}	Red
4 → 2	4.861×10^{-5}	4.8613×10^{-5}	Blue-green
5 → 2	4.340×10^{-5}	4.3405×10^{-5}	Blue
4 → 3	18.75×10^{-5}	18.751×10^{-5}	Infrared

The almost exact agreement between wavelengths as predicted by Bohr and observed wavelengths is very gratifying.

Prior to Bohr, there had already been several partially successful attempts at explaining the spectrum of the hydrogen atom without invoking the quantum theory. While those theories could predict qualitatively that the spectrum should be a line spectrum, none of them succeeded in reproducing the actual wavelengths.

Fig. 9-5
De-excitation of an excited hydrogen atom.

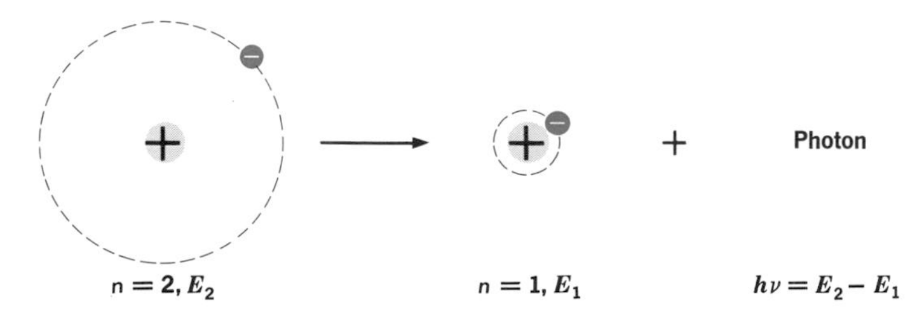

Bohr's precise calculation of the spectrum of the hydrogen atom was of course a spectacular success. But his even greater achievement was to introduce the concepts of quantum theory into chemistry and to apply them to an atomic model that is consistent with Rutherford's experiments. This approach has proved enormously fruitful and has led to all of modern atomic theory. As time passed, Bohr's theory was supplanted by the more general theory of wave mechanics (Ch. 11). But this does not detract from its historical importance, for this theory represents the point of departure from classical Newtonian ideas about the behavior of atoms and replaces them with ideas that have a particular applicability to the submicroscopic world alone.

Suggestions for further reading

Butterfield, H., "The Scientific Revolution," *Scientific American,* Sept. 1960, p. 173.

Feinberg, G., "Light," *Scientific American,* Sept. 1968, p. 50.

Musgrave, T. R., "The Incredible Shrinking Universe Theory," *Journal of Chemical Education, 46,* Sept. 1969, p. 546.

Nelson, D. F. "The Modulation of Laser Light," *Scientific American,* June 1968, p. 17.

Oppenheimer, J. R., "The Age of Science: 1900–1950," *Scientific American,* Sept. 1950, p. 20.

Rush, J. H., "The Speed of Light," *Scientific American,* Aug. 1955, p. 62.

Schrödinger, E., "What is Matter?" *Scientific American,* Sept. 1953, p. 52.

Szabadvary, F., "From Thales to Bohr," *Chemistry, 42,* Dec. 1969, p. 6.

Tarski, A., "Truth and Proof," *Scientific American,* June 1969, p. 40.

Questions

1. The electromagnetic spectrum, in order of decreasing wavelength, extends from radio waves to X-rays. List the names and typical wavelengths of intermediate regions in the electromagnetic spectrum.

2. (*a*) Do all electromagnetic waves travel at the same speed in a vacuum? (*b*) Do light waves travel at the same speed in water as in air? Do beams of red and blue light travel at the same speed in these media?

3. (*a*) Calculate the frequency of a beam of orange light whose wavelength (in a vacuum) is 6.00×10^{-5} cm. (*b*) Find the wavelength of radio waves whose frequency is 60,000,000 $sec.^{-1}$ (read: sixty million per second).

4. Which wire is hotter, the one at a "red heat" or the one at a "white heat"?

5.* The pendulum in a grandfather clock provides a familiar example of a macroscopic oscillator. The frequency of each pendulum is rigidly fixed—in many cases at two swings per second. The energy of the swinging pendulum is continuously variable, increasing with the amplitude of the swinging motion. Contrast these properties with those of the submicroscopic oscillators envisaged by Planck.

6. What is the energy of one photon of green light whose wavelength (in a vacuum) is 5.46×10^{-5} cm.? What is the energy of one mole (Avogadro's number) of such photons?

7. According to Einstein's theory of the photoelectric effect, how many photons (of the appropriate energy) are required to cause the emission of 100 electrons?

8. In Rutherford's model of the hydrogen atom, a negatively charged electron of charge -1 moves in a stable orbit around a much heavier nucleus of charge $+1$. Explain why such an atom possesses both kinetic and potential energy. What is the algebraic sign (positive or negative) of the kinetic energy? Of the potential energy?

9. Define or explain:
(*a*) Ground state of the hydrogen atom.
(*b*) First excited state of the hydrogen atom.
(*c*) Transition of the atom from the first excited state to the ground state with emission of a photon.
(*d*) Quantum number.

10. According to Bohr's theory, as the quantum number increases, the radius of the electron's orbit increases while the velocity and frequency of its circular motion decreases. Compare these properties with those of a man-made satellite moving in a circular orbit around the earth.

11. The electron orbits envisaged by Bohr are planar orbits. Does Bohr's theory therefore imply that the hydrogen atom is flat? (Consider the effect to an outside observer if the entire atom is spinning.)

12. Calculate the wavelength of the light emitted when hydrogen atoms change from $n = 3$ to $n = 1$. Calculate the energy of the emitted photon.

13. Seen in historical perspective, what was Bohr's most important contribution to the development of atomic theory?

14. What are the principle differences between Bohr's theory and Rutherford's theory of the hydrogen atom?

* Questions marked with an asterisk (*) are more difficult or require supplementary reading.

The scientist must set in order. Science is built up with facts, as a house is with stones. But a collection of facts is no more a science than a heap of stones is a house. And above all, the scientist must foresee.

HENRI POINCARÉ (1854–1912)

Ten

FAMILIES OF ELEMENTS AND THE PERIODIC CHART

BEFORE CONSIDERING the electronic structure of atoms that are more complicated than hydrogen, we need to have additional information about their chemical properties, in order to have an experimental basis for any theories that might be developed. When we examine the properties of the 105 chemical elements, we find a seemingly bewildering array of reactions and compounds. However, we can bring some order to this chaotic state of affairs by classifying the elements according to the types of reaction they undergo and the formulas of the compounds that are obtained.

To begin with, we note that there are two broad classes of elements: metals and nonmetals. The metals do not usually combine with other metals, but form compounds readily with nonmetals. On the other hand, the nonmetals react not only with metals, but also with other nonmetals.

The metals can be distinguished easily from the nonmetals by their physical properties: freshly cut metal surfaces have a characteristic luster, and metals are good conductors of heat and electricity. Well-known metallic elements are iron, nickel, chromium, copper, tin, aluminum, magnesium, silver, gold, and platinum. These are all solids at room temperature. A well-known metal that is liquid at room temperature is mercury or *quicksilver*, which is commonly used by dentists for making silver amalgam, and in clinical thermometers. Less well known are such reactive metals as sodium, potassium, and calcium. These corrode too quickly for use in the open air.

The nonmetals include the gases of the atmosphere: nitrogen, oxygen, and argon. Other nonmetals that we have already mentioned are chlorine, bromine, sulfur, carbon, and phosphorus. The nonmetals are poor conductors of heat and electricity; and when they are in the solid state, they are usually brittle.

Families of elements

It was noted as early as 1820 that certain of the elements can be classified into *families* on the basis of their very similar chemical properties. For example, the nonmetals chlorine and bromine both belong to a family known as the *halogens*.† Not only do chlorine and bromine react with the same substances, but the molecular formulas of the reaction products are analogous. Both react with sodium to produce the compounds NaCl and NaBr; both form similar compounds with carbon, CCl_4 and CBr_4; both react with mercury to form not one but two different compounds: $HgCl_2$ and Hg_2Cl_2, $HgBr_2$ and Hg_2Br_2; and both react with hydrogen to form HCl and HBr. The list can be extended almost indefinitely. On the other hand, few cases are known in which one of these halogens reacts and the other does not. Therefore it can be stated as an almost general rule that when chlorine reacts with a given substance, so does bromine; and we can predict the molecular formula of the bromine compound from that of the chlorine compound by merely writing the symbols for bromine atoms instead of chlorine atoms.

This generalization is enormously useful. Suppose that chlorine reacts with metallic calcium to form the compound calcium chloride, $CaCl_2$. Then we can predict with reasonable certainty that bromine will also react with calcium, and that the molecular formula of the compound will be $CaBr_2$—not CaBr, or $CaBr_3$, or Ca_2Br, but specifically $CaBr_2$

For contrast, let us consider two elements that *do not* belong to the same family, such as chlorine and nitrogen. Both elements can be made to react with hydrogen. Whereas chlorine forms only a single stable compound whose formula is HCl, nitrogen forms a number of stable

† From a Greek root meaning "salt producers."

compounds, among them NH_3 (ammonia), HN_3 (hydrazoic acid), and N_2H_4 (hydrazine), but not the analogous compound HN. Chlorine reacts with sodium to form the very stable salt sodium chloride, NaCl, whereas the most stable compound of sodium and nitrogen, sodium azide, has the formula NaN_3. Chlorine reacts with phosphorous to form PCl_3 and PCl_5, but no stable compound containing only phosphorous and nitrogen is known.

These remarks may be generalized as follows:

1. When one of the elements in a family reacts (or fails to react) with a given substance, then all the others will probably behave in a similar manner.
2. The molecular formulas of the compounds formed from the elements of a given family are similar.
3. When a *compound* of one of the elements of the family reacts with a given substance, then the analogous compounds of the other elements in the family will probably react also; and the molecular formulas of the reaction products will be similar.

The existence of families of elements greatly simplifies the task of learning chemistry. If we know the reactions of just one member of each family and the names of the other members, then we can predict by analogy the reactions of all the members of all the families. Thus, it is possible for an ordinary person to memorize enough selected facts so that he can predict a rather impressive amount of chemistry.

It turns out that about half of the elements can be classified into eight well-defined families. The remaining elements, which are not included in these eight families, are all metallic in character. In the next few pages we shall give a short summary of four of the well-defined families of elements.

THE ALKALI METALS

The so-called *alkalis*, potash and soda, have been known since ancient times. Soda (sodium carbonate) was gained primarily from mineral sources, e.g., it could be leached from small alkaline hillocks in the fields around Smyrna and Ephesus in Asia Minor. The detergent property of this alkali is mentioned even in the Bible.† Potash (potassium carbonate) was obtained primarily from the ash of certain plants. These substances have long been important raw materials in the manufacture of glass.

In 1807, Sir Humphry Davy succeeded in preparing two pure alkali metals—which he named *potassium* (from potash) and *sodium* (from soda)—by the electrolysis‡ of the slightly moist alkalis. The metals proved to be extremely reactive; they corroded rapidly in air and gave a vigorous reaction with water. The other alkali metals were discovered somewhat later. The element lithium was discovered by the Swedish chemist, J. H. Arfwedson, in 1817, and the pure metal was prepared in sufficient quantity for the study of its physical properties by the Germans, R. W. Bunsen and A. Mathiessen, in 1855. The elements cesium and rubidium were discovered by Bunsen and G. R. Kirchhoff in 1860–61. The element francium was first isolated in tangible amounts in 1944 by scientists of the United States Manhattan Project. It is a fission product of uranium (p. 339). The alkali metals and some of their properties are shown in the table.

THE ALKALI METALS

Name	Symbol	Atomic number	Atomic weight	Melting point (C)	Boiling point (C)	Atomic radius (cm)
Lithium	Li	3	6.94	186°	1,336°	1.55×10^{-8}
Sodium	Na	11	22.99	97.5°	880°	1.90×10^{-8}
Potassium	K	19	39.10	62.3°	760°	2.35×10^{-8}
Rubidium	Rb	37	85.47	38.5°	700°	2.48×10^{-8}
Cesium	Cs	55	132.91	28.5°	670°	2.67×10^{-8}
Francium	Fr	87	[223]			

The reader will note that the physical properties vary in a regular manner as the atomic weight of the element increases. A regular variation of properties with increasing atomic weight is not uncommon for the elements of a given family and may be used to predict approximate values for unknown properties. For example, if the melting point of rubidium

† "For though thou wash thee with nitre [archaic term for sodium carbonate] and take thee much sope, yet thine iniquity is marked before me." Jeremiah 2:22.

‡ See p. 214.

were not already known, one might have guessed that it must lie between the melting point of potassium and that of cesium, probably in the neighborhood of 45 °C, in fair agreement with the actual value.

The formulas and names of typical compounds of the alkali metals are exemplified by some compounds of sodium:

$NaCl$	Sodium chloride (common table salt)
$NaOH$	Sodium hydroxide (caustic soda)
Na_2CO_3	Sodium carbonate (soda)
$NaHCO_3$	Sodium hydrogen carbonate (baking soda, bicarbonate of soda)
Na_2O	Sodium oxide

The other alkali metals form analogous compounds; for example, compounds of the formula $LiCl$, Rb_2CO_3, KOH are all known to be stable.

It is customary that new elements be named by their discoverers, although the name is eventually approved by an official international organization such as the International Union of Chemistry. An element is not officially "discovered" until the proof of its existence is such as to satisfy even the most conservative members of the profession. In the nineteenth century it was considered necessary to isolate enough of the new element so that its atomic weight and a few of its properties could be determined. Thus, to prove the existence of one of the less abundant elements, it was necessary to subject hundreds of pounds of ore to chemical separation procedures in order to isolate a precious gram or so of the new element. Nowadays this requirement is regarded as too stringent because all of the stable, naturally occurring elements have by now been discovered. Any new elements must be prepared artificially, and since such elements are all radioactive, they exist only for a limited time. In the case of the element mendelevium (atomic number 101), discovery was officially acknowledged after only a hundred or so *atoms* had been prepared.

At the present time (1970), the International Union of Chemistry is in the process of approving names for two recently-discovered elements, with atomic numbers 104 and 105. The Union will have to decide between U. S. and Russian claims for prior discovery, and then will probably approve the name suggested by the discoverers. A consensus seems to be developing that element 104 should be named kurchatovium, in honor of the Russian, Igor Kurchatov, and that element 105 should be named hahnium, in honor of Otto Hahn, the German co-discoverer of nuclear fission.

There are few systematic rules concerning the names of the elements. The names of metals end in *-um* or *-ium,* except for certain metals like copper, iron, and tin, which have been known since ancient times. Potassium and sodium were named after the materials from which they were obtained. Lithium was named after a Greek root meaning "stone" to acknowledge its isolation from mineral sources. Rubidium (meaning red) and cesium (meaning blue) were named after the color of the light these elements emit when they are in the gaseous state at a high temperature. Some elements are named in honor of countries, like germanium for Germany or scandium for Scandinavia; others in honor of cities, like berkelium for Berkeley. Still others are named in honor of people, like curium for the investigators of radioactivity, Pierre and Marie Curie. Occasionally elements are named for mythological figures such as promethium for Prometheus who, according to the myth, stole fire from the gods for the benefit of mankind and was sentenced for this crime to eternal agony.

THE ALKALINE EARTH METALS

Several compounds with properties similar to those of potash and soda have been known for centuries, but the metals obtained from them clearly are not alkali metals, but belong to another family and are called *alkaline earth metals*. The alkaline earth metals are less reactive than the alkali metals, have higher melting points, and are harder and stronger. The metals and some of their properties are shown in the table.

THE ALKALINE EARTH METALS

Name	Symbol	Atomic number	Atomic weight	Melting point (C)	Boiling point (C)	Atomic radius (cm)
Beryllium	Be	4	9.01	1,350°	about 1,600°	1.12×10^{-8}
Magnesium	Mg	12	24.31	651°	1,110°	1.60×10^{-8}
Calcium	Ca	20	40.08	810°	1,170°	1.97×10^{-8}
Strontium	Sr	38	87.62	800°	1,150°	2.15×10^{-8}
Barium	Ba	56	137.34	850°	1,140°	2.22×10^{-8}
Radium	Ra	88	226.03	960°	1,140°	. . .

Magnesium, calcium, strontium, and barium were first prepared by Sir Humphry Davy in 1808. Beryllium was isolated in 1828 independently by F. Wöhler and A. A. B. Bussy. The radioactive element radium was discovered by Pierre and Marie Curie in 1898.

The formulas and names of some typical compounds of the alkaline earth metals are illustrated by the following compounds of calcium:

$CaCl_2$	Calcium chloride
$CaCO_3$	Calcium carbonate—marble consists of calcium carbonate, and limestone is a rock composed mainly of this substance
CaO	Calcium oxide (quicklime)
$Ca(OH)_2$	Calcium hydroxide (slaked lime)
$CaSO_4$	Calcium sulfate—gypsum and plaster of Paris are compounds consisting of calcium sulfate and water
CaC_2	Calcium carbide—obtained when calcium oxide is heated with coke (carbon)

The other alkaline earth metals form analogous compounds; for example, MgO (magnesium oxide or magnesia), $BaSO_4$, $BeCl_2$, and $SrCO_3$ are known to be the formulas of stable compounds.

THE HALOGENS

The halogens are among the most active of the nonmetals and will form compounds with all but a few of the other elements. A list of the halogens and of some of their properties is given in the table.

THE HALOGENS

Name	Symbol	Atomic number	Atomic weight	Melting point (C)	Boiling point (C)	Radius of free atom (cm)
Fluorine	F	9	19.00	−223°	−187°	1.35×10^{-8}
Chlorine	Cl	17	35.45	−101.6°	−34.6°	1.80×10^{-8}
Bromine	Br	35	79.90	−7.3°	58.7°	1.95×10^{-8}
Iodine	I	53	126.90	113.5°	184°	2.15×10^{-8}

Fluorine is a pale yellow gas at room temperature; chlorine is a greenish-yellow gas; bromine is a dark red liquid; and iodine is a greyish-black solid. Chlorine was first prepared in 1774 by C. W. Scheele in

Sweden, and it was shown to be an element by Sir Humphry Davy in 1807. Iodine was discovered by B. Courtois soon thereafter (in 1811), and bromine was discovered in 1826 independently by A. J. Balard and by Carl Löwig. Both men were young students at the time. The discovery of fluorine took place much later because this gas is extremely reactive and therefore dangerous to handle and difficult to prepare. It is also very poisonous, and at least two competent chemists are known to have died while attempting to prepare it. Fluorine was finally obtained in 1886 by the French chemist, Henri Moissan, from the electrolysis of potassium fluoride dissolved in hydrogen fluoride at a low temperature, in apparatus made entirely of the precious metal platinum. The use of platinum was necessary because hydrogen fluoride will dissolve glass. If fluorine were not so difficult to handle, it would make an outstanding rocket fuel because of the large amount of energy released when it reacts with other substances.

The formulas and names of some of the compounds of the halogens are illustrated by the following compounds of chlorine:

$CsCl$	Cesium chloride—this salt is sometimes used to make prisms and windows for certain optical instruments
$AlCl_3$	Aluminum chloride—an important chemical catalyst
$SnCl_4$	Stannic chloride
PCl_3	Phosphorous trichloride
PCl_5	Phosphorous pentachloride
S_2Cl_2	Sulfur monochloride
ICl	Iodine chloride

The reactivity of the halogens usually decreases in the order fluorine, chlorine, bromine, iodine. The reactivity of compounds of the halogens often varies in the opposite order. For example, the compound CH_3Cl is less reactive than CH_3Br.

THE NOBLE GASES

There is a family of elements, the *noble gases*, which are so unreactive that no stable compounds of these elements were known for more than half a century after their discovery. The names and properties of the noble gases are shown in the table.

THE NOBLE GASES

Name	Symbol	Atomic number	Atomic weight	Melting point (C)	Boiling point (C)	Atomic radius (cm)
Helium	He	2	4.003	– 272°	– 268.9°	0.8×10^{-8}
Neon	Ne	10	20.18	– 248.67°	– 245.9°	1.1×10^{-8}
Argon	Ar	18	39.95	– 189.2°	– 185.7°	1.5×10^{-8}
Krypton	Kr	36	83.80	– 157°	– 152.9°	1.7×10^{-8}
Xenon	Xe	54	131.30	– 112°	– 107.1°	1.9×10^{-8}
Radon	Rn	86	222	– 71°	– 61.8°	. . .

The first noble gas to be discovered was argon, a constituent of the atmosphere making up about 1% of ordinary air. The discovery took place in 1894, fairly late in the history of the discovery of the stable elements, and the credit for it goes to Lord Rayleigh and Sir William Ramsay. The story of the discovery of argon is of interest because it is a good example of research that was free to investigate unexpected and startling phenomena when they turned up. Lord Rayleigh was investigating the atomic weight of nitrogen and, in that connection, prepared nitrogen gas by three different methods. One of the methods he used was to separate nitrogen from the other components of the atmosphere. He subjected ordinary air to various chemical procedures until he ended up with a gas that should have been pure nitrogen. But to his consternation, the density of the gas was just slightly different from that of samples prepared by other methods, one of which was the decomposition of ammonia. Guided by some experiments carried out by the British scientist, Henry Cavendish, more than a hundred years earlier, Rayleigh suspected that the "nitrogen" obtained from ordinary air still contained an impurity. Collaborating with Ramsay, he subjected the gas to all sorts of chemical procedures that were designed to remove any and all conceivable impurities. When all of these failed, Ramsay decided to abandon the original investigation and to study the impurity. He passed the impure nitrogen over hot magnesium, which removed the nitrogen gas by forming solid magnesium nitride, Mg_3N_2. A small amount of gas was left unreacted which had physical properties unlike those of any known gas. It proved to be a new element. Because of its lack of reactivity, the new gas was named *argon,* which is Greek for "the lazy one."

The other noble gases, with the exception of radon, were discovered by Ramsay and H. W. Travers soon thereafter. Radon, the product of the radioactive decay of radium, was, of course, not discovered until after the isolation of substantial amounts of radium had been accomplished. Radon itself is a radioactive gas and decays within a matter of days to form another radioactive element, polonium.

The molecules of the noble gases were found to be monatomic, that is, they consist of uncombined atoms. These atoms are evidently quite stable and have little tendency to enter into chemical combination. As recently as 1961, no compounds of the inert gases had ever been isolated. However, with modern advances in chemical apparatus and technology, and with the availability of highly energy-rich reagents, chemists finally succeeded, in 1962, in forcing the atoms of xenon and krypton into chemical combination. The compounds that have been prepared are fluorides and oxides. Substances with the formulas XeF_2, XeF_4, XeF_6, $XeOF_4$, XeO_3, and KrF_4 are definitely known, and research in this field is continuing. So far, no compounds have as yet been prepared from helium, neon, and argon.

The discovery of the noble gases has had far-reaching effects on chemical theory. Since the noble gas atoms are exceptionally stable in the uncombined state, they must have little to "gain" in the way of additional stability by combining with other atoms. Thus, by studying the electronic structure of these atoms, we may hope to discover a clue to the type of electronic structure that other, less stable, atoms acquire when they combine to form compounds.

HYDROGEN

The element hydrogen enjoys a unique status among the elements. Some of its reactions resemble those of the alkali metals, and others resemble those of the halogens. For example, hydrogen reacts with chlorine to form hydrogen chloride (HCl, compare with $NaCl$) and with sulfur to form hydrogen sulfide (H_2S, compare with Na_2S). But hydrogen also reacts with sodium to form sodium hydride (NaH, compare with $NaCl$) and with calcium to form calcium hydride (CaH_2, compare with $CaCl_2$).

Some of the physical properties of hydrogen are as follows:

Atomic number:	1	Melting point:	− 259 °C
Atomic weight:	1.008	Boiling point:	− 252.7 °C

At room temperature, hydrogen is a colorless and odorless gas. Hydrogen was first distinguished from other gases by Henry Cavendish around 1770.

The periodic chart of the elements

In Dalton's atomic theory, the most important property which distinguishes the atoms of one element from those of another is the atomic weight. Early in the development of chemistry, attempts were therefore made to find regular relationships between the atomic weights of the elements and their chemical properties. The pioneer in this field was the German chemist, J. W. Döbereiner, who made notable progress as early as 1820. However, the reader will recall that at that time, the atomic weight scale had not yet been firmly settled, and that only relatively few of the elements were known. Chemists of the time were busy debating atomic theory and laying the foundations of organic chemistry (the chemistry of carbon compounds). They were not ready to move on this "front" as well.

Forty years later, after the question of the atomic weight scale had been solved and many new elements had been discovered, the time was ripe for a new search for relationships between atomic weights and chemical properties. Many celebrated chemists participated in that search, notably J. A. R. Newlands of Britain and Lothar Meyer of Germany. Yet the major credit for the final result of their efforts—the periodic chart of the elements—goes to the Russian chemist, Dimitri I. Mendeleev.

After the scientific revolution at the beginning of the twentieth century, it was realized that the distinguishing property of each element is not its atomic weight, but its nuclear charge, expressed conveniently by means of the atomic number (see Ch. 8). It is therefore logical to try to relate the chemical properties of the elements to their atomic number. Atomic numbers were not available to Mendeleev, however, who there-

Dimitri Ivanovitch Mendeleev: 1834–1907
Mendeleev's periodic chart, pubished in 1869, systematized the properties of the elements and predicted the existence of several then undiscovered elements. [Photograph courtesy Culver Service, Inc.]

fore used atomic weights. His results, nevertheless, are in substantial agreement with the results one obtains using atomic numbers.

If the first twenty elements are arranged in order of increasing atomic number, as follows,

H He Li Be B C N O | F Ne Na Mg Al Si P S | Cl Ar K Ca

there is a similarity between every ninth element. Therefore, if these elements are arranged in groups of eight, and if the groups are placed under each other in successive rows as follows,

H	He	Li	Be	B	C	N	O
F	Ne	Na	Mg	Al	Si	P	S
Cl	Ar	K	Ca				

then the elements standing together in the same vertical columns are members of the same chemical family. It will be noted that hydrogen in this listing stands with the halogens, although it has many unique properties. See page 168.

For elements of higher atomic number, the situation becomes more complex. Nevertheless it is possible to organize all elements into a single periodic arrangement, as shown in Fig. 10-1, which is called a *periodic chart of the elements*.

Figure 10-1 arranges the elements in rows or *periods* such that each

PERIODIC CHART OF THE ELEMENTS

Row	I	II											III	IV	V	VI	VII	VIII
—	1 H Hydrogen												Metalloids and Non-metals					2 He Helium
First	3 Li Lithium	4 Be Beryllium						METALS					5 B Boron	6 C Carbon	7 N Nitrogen	8 O Oxygen	9 F Fluorine	10 Ne Neon
Second	11 Na Sodium	12 Mg Magnesium					Transition Metals						13 Al Aluminum	14 Si Silicon	15 P Phosphorus	16 S Sulfur	17 Cl Chlorine	18 Ar Argon
Third	19 K Potassium	20 Ca Calcium	21 Sc Scandium	22 Ti Titanium	23 V Vanadium	24 Cr Chromium	25 Mn Manganese	26 Fe Iron	27 Co Cobalt	28 Ni Nickel	29 Cu Copper	30 Zn Zinc	31 Ga Gallium	32 Ge Germanium	33 As Arsenic	34 Se Selenium	35 Br Bromine	36 Kr Krypton
Fourth	37 Rb Rubidium	38 Sr Strontium	39 Y Yttrium	40 Zr Zirconium	41 Nb Niobium	42 Mo Molybdenum	43 Tc Technetium	44 Ru Ruthenium	45 Rh Rhodium	46 Pd Palladium	47 Ag Silver	48 Cd Cadmium	49 In Indium	50 Sn Tin	51 Sb Antimony	52 Te Tellurium	53 I Iodine	54 Xe Xenon
Fifth	55 Cs Cesium	56 Ba Barium	57 La Lanthanum	72 Hf Hafnium	73 Ta Tantalum	74 W Tungsten	75 Re Rhenium	76 Os Osmium	77 Ir Iridium	78 Pt Platinum	79 Au Gold	80 Hg Mercury	81 Tl Thallium	82 Pb Lead	83 Bi Bismuth	84 Po Polonium	85 At Astatine	86 Rn Radon
Sixth	87 Fr Francium	88 Ra Radium	89 Ac Actinium	104	105													
	Lanthanides (Rare Earth Metals)				58 Ce Cerium	59 Pr Praseodymium	60 Nd Neodymium	61 Pm Promethium	62 Sm Samarium	63 Eu Europium	64 Gd Gadolinium	65 Tb Terbium	66 Dy Dysprosium	67 Ho Holmium	68 Er Erbium	69 Tm Thulium	70 Yb Ytterbium	71 Lu Lutetium
	Actinides				90 Th Thorium	91 Pa Protactinium	92 U Uranium	93 Np Neptunium	94 Pu Plutonium	95 Am Americium	96 Cm Curium	97 Bk Berkelium	98 Cf Californium	99 Es Einsteinium	100 Fm Fermium	101 Md Mendelevium	102 No Nobelium	103 Lw Lawrencium

FIG. 10–1.

row begins with an alkali metal and ends with a noble gas. (This ignores the first two elements, hydrogen and helium.) The chart thus places all of the alkali metals into the single vertical column labeled I, and all of the noble gases appear in the single column labeled VIII. In this way, one achieves the remarkable result that all elements appearing in the same column are members of one and the same family. For example, column II lists all the alkaline earth metals, and column VII all the halogens. The elements shown in the other columns have not yet been discussed in detail, but here again, each column contains all the elements belonging to that particular family. For example, in column III there is a strong family resemblance between boron (B) and aluminum (Al), as shown by the similar molecular formulas of some of their compounds, such as $AlCl_3$ and BCl_3, or $LiBH_4$ and $LiAlH_4$.

A number of columns appear under the single heading, Transition Metals. This group of elements includes some of the best known metals of our civilization, such as iron, copper, zinc, nickel, and silver. Here again, each column corresponds to a family, but the transition elements are enough alike so that it is convenient to talk about them as if they were a single large family.

Closer examination of the periodic chart shows that there is an irregularity beginning with element 58 and ending with element 71, and also beginning with element 90 and continuing to the end of the chart. While elements 57 and 89 resemble 21 and 39, element 58 is *not* a member of the same family as 22 and 40, element 59 is *not* a member of the same family as 23 and 41, and so on. It is not until we come to element 72 that we have again a member of the family to which 22 and 40 belong. Elements 58 through 71 are almost identical chemically and therefore form a family all by themselves. This is the family of the lanthanides or *rare-earth metals,* so named because of their rare occurrence in nature. Similarly, the elements beginning with element 90 and continuing to the end of the chart form a separate family, called the *actinide* family. Since these elements form separate families, they have been placed in a separate portion of the chart.

A dividing line has been drawn in the periodic chart to separate the metals from the nonmetals. Actually, the distinction between metals and nonmetals is not always clear-cut, and some of the elements bordering on the dividing line have properties that are intermediate between those of metals and nonmetals. For example, the electrical conductivity of the borderline element germanium is a thousand times smaller than that of tin, a more typical metal, but it is enormously larger than that of white phosphorous, a typical nonmetal. This borderline conductivity has had enormous practical consequences. Germanium, along with a number of other substances, such as silicon, are *semiconductors*. These form the basis of the electronic *transistor* which has made possible the high speed computer and miniaturized electronic components essential to the success of the program of space exploration.

Periodic behavior of the elements is evident not only in their chemical properties, but also in certain physical properties. See Figs. 10-2 to 10-4.

Fig. 10-2
Ionization energy in volts, of the first sixty elements. The ionization energy is the work required to remove an electron from the uncombined atom in the gas phase.

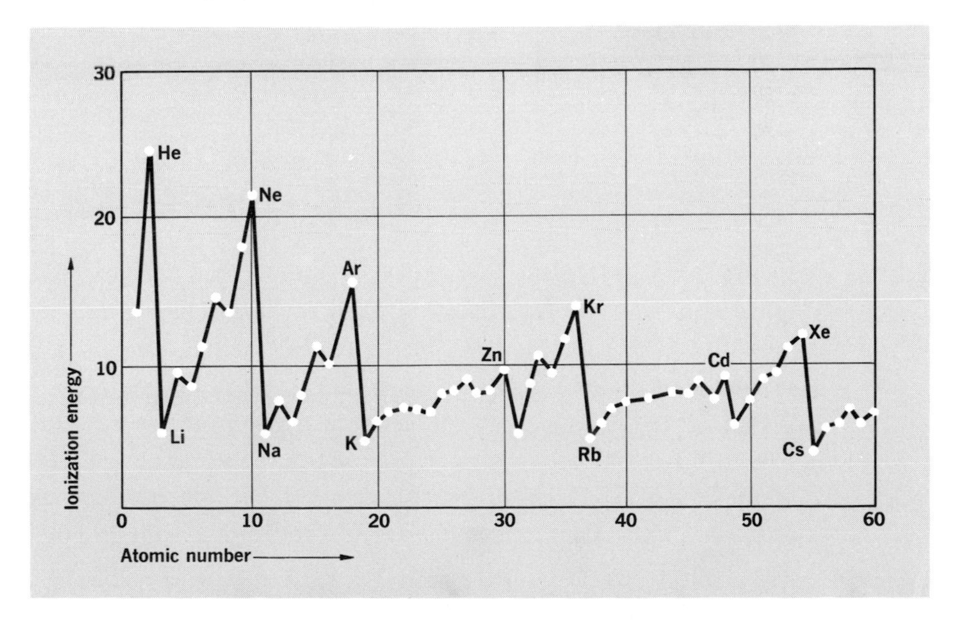

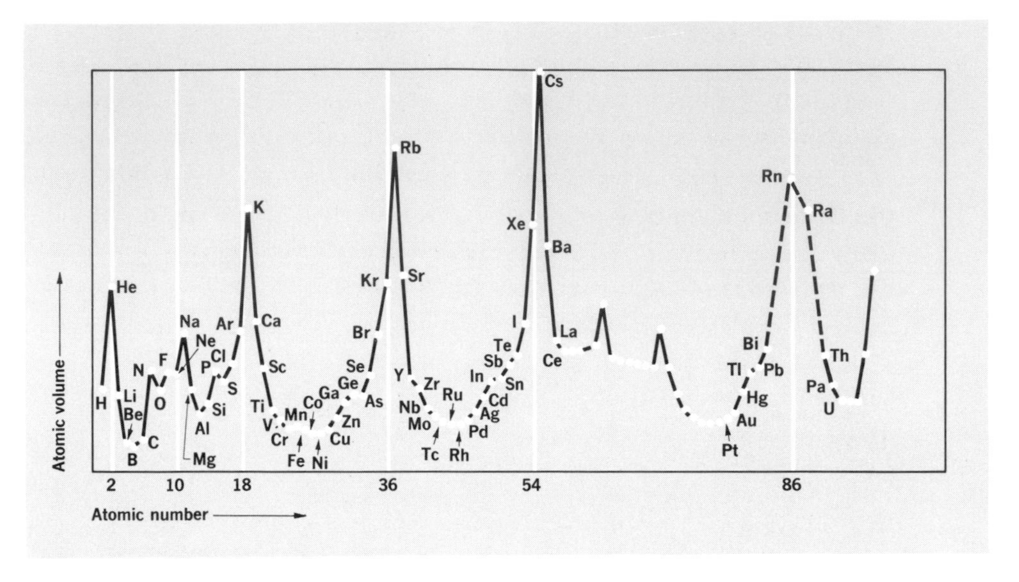

Fig. 10-3
Variation of atomic volume of the elements with atomic number. The atomic volume is taken as the volume of one gram atom of the pure element in the solid state at a given reference temperature. Vertical lines indicate atomic numbers of inert gases.

Uses of the periodic chart

In 1889, D. I. Mendeleev was guest of honor of the Chemical Society of London in recognition of his work on the formulation and use of the periodic chart. The address he delivered at that time is a beautiful summary of the various uses to which the periodic chart can be put.† We shall describe some of those uses in this section. In order to gain the correct historical perspective, we must bear in mind that 1889 was still in the era of Dalton's unstructured atom; atomic numbers were still unknown, and several elements were as yet undiscovered. The periodic

† D. I. Mendeleev, *Journal of the Chemical Society* (London), *55*, (1889) pp. 634–656.

chart of Mendeleev was obtained by listing the elements in order of increasing atomic weight, and since the atomic weight usually increases as the atomic number increases, his chart was essentially the same as Fig. 10-1.

Perhaps the most important feature of the periodic chart is that all the elements belonging to a given family are listed in a single column, and the similarities of their chemical reactions are thereby emphasized. Thus, new reactions and the molecular formulas of new compounds can readily be predicted from the analogy to known facts about the other members of the same family.

However, there are also some striking regularities among the elements in any one period or row. For example, the following regularities have been noted for the series of elements silver (Ag), cadmium (Cd), indium (In), tin (Sn), antimony (Sb), tellurium (Te), and iodine (I), which occupy adjacent positions in the fourth row. Note the regular decrease in density, the regular change in the formula of the oxide.

Fig. 10-4
Atomic radii of the elements versus the atomic number.

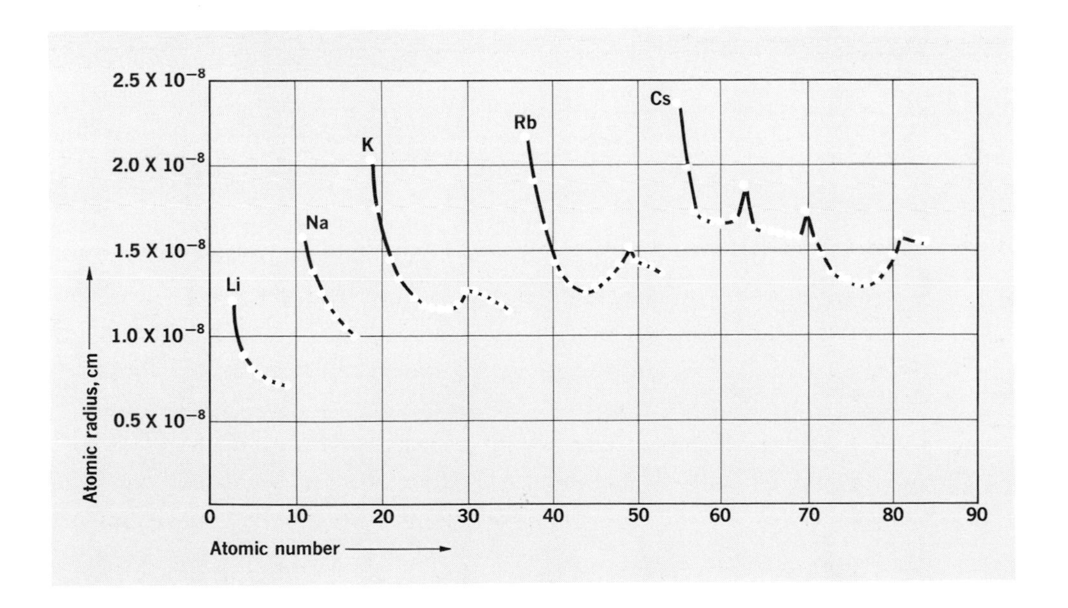

COMPARISON OF FOURTH-ROW ELEMENTS

Element	Ag	Cd	In	Sn	Sb	Te	I
Density of pure element	10.5	8.6	7.4	7.2	6.7	6.4	4.9
Formula of oxide†	Ag_2O	Cd_2O_2	In_2O_3	Sn_2O_4	Sb_2O_5	Te_2O_6	I_2O_7

† Formula as written by Mendeleev, to emphasize the regular increase in the quantity of oxygen in the series.

Perhaps the most important use of the periodic chart was to pinpoint the as-yet-undiscovered elements. Before the formulation of the chart, there was no special reason to expect the discovery of new elements, and the new ones that were discovered from time to time appeared to be possessed of quite novel properties. The periodic chart changed all that. For example, in 1871 no elements were known whose properties would fit the properties to be expected for the two elements between zinc (Zn) and arsenic (As) in the third row of the periodic chart. Mendeleev therefore concluded that these elements had not yet been discovered and left two empty spaces in the chart. He even went so far as to predict the properties of the elements that would eventually occupy these spaces, solely on the basis of the properties of the other known elements in the same column or row. Within less than two decades both elements had been discovered, and Mendeleev lived to see his predictions confirmed with uncanny accuracy. For example, the predicted and actual properties of the element germanium (discovered by the German chemist, C. A. Winkler, 1886) are shown below:

PREDICTIONS BASED ON THE PERIODIC CHART

Property of germanium	Value predicted by Mendeleev	Actual value
Atomic weight	72	72.59
Density of element	5.5	5.36
Formula of the oxide	GeO_2	GeO_2
Density of the oxide	4.7	4.70
Formula and properties of the chloride	$GeCl_4$, boiling point a little under 100°C, density 1.9	$GeCl_4$, boiling point 83°C, density 1.88

Another use of the periodic chart was to indicate possible errors in the reported properties of known elements. For example, the element beryllium, whose chemistry was known only incompletely in 1871, had tentatively been assigned to the boron family. But Mendeleev realized that there was no room for another element of low molecular weight in column III, whereas there was an empty space at the head of Column II. He therefore concluded that beryllium must be an alkaline earth element, and later research has fully confirmed that view.

IS THE PERIODIC CHART COMPLETE?

The various horizontal rows (periods) of the periodic chart differ in the number of elements that each contains. After hydrogen and helium, the first two rows consist of eight elements each, the third and fourth of eighteen elements each, and the fifth of thirty-two elements. The sixth row contains nineteen elements and is presumed to be incomplete.

Since the number of the elements in each of the rows will be an important clue in the explanation of the chart, we must ask ourselves whether we can be sure that these numbers are right. In other words, are we safe in assuming that all the elements that properly belong in the first five rows have by now been discovered? The answer to this question is decidedly "yes," and for the following reason: the atomic number of an element is equal to the number of electrons in the neutral atom. Since there is no such thing as half an electron or a fraction of an electron, the atomic number must always be a whole number. We can therefore be sure that we know all the elements in a given row if we can list an element for each whole number. For example, in the second row there is a large empty space between magnesium and aluminum, and we might wonder whether any undiscovered elements should go in there. But since the atomic number of magnesium is twelve and that of aluminum is thirteen, we can safely say that the space between magnesium and aluminum should be left empty.

We have now examined a substantial body of experimental data and its organization by means of the periodic chart. This organization strongly implies that there is a corresponding regularity in the submicroscopic structure of the atoms. In the next chapter, we shall inquire whether an extension of Bohr's model of the hydrogen atom can explain the regularity.

Suggestions for further reading

Bartlett, N., "The Chemistry of the Noble Gases," *Endeavor, 23,* January 1964, p. 3.

Conant, J. B., "Theodore William Richards and the Periodic Table," *Science, 168,* 24 April, 1970, p. 425.

Cook, N. C., "Metalliding," *Scientific American,* August 1969, p. 38.

Massey, A. G., "Boron," *Scientific American,* January 1964, p. 88.

Mellor, D. P., "The Noble Gases and Their Compounds," *Chemistry, 41,* October 1968, p. 16.

Seaborg, G. T., "Prospects For Considerable Expansion of the Periodic Table," *Journal of Chemical Education, 46,* October 1969, p. 626.

Selig, H., Malm, J. G., and Claassen, H. H., "The Chemistry of the Noble Gases," *Scientific American,* May 1964, p. 66.

Tilden, W. A., "Mendeleev Memorial Lecture," *Journal of the Chemical Society,* (London), *95* 1909, p. 2077.

van Spronsen, J. W., "The Priority Conflict Between Mendeleev and Meyer," *Journal of Chemical Education, 46,* March 1969, p. 136.

Williams, L. P., "Humphry Davy," *Scientific American,* June 1960, p. 106.

"Enter No. 105", *Newsweek,* May 11, 1970, p. 109.

"Element 105 is Long-lived," *Chemical and Engineering News,* May 4, 1970, p. 9.

Questions

1. Name three properties of the elements that vary periodically with atomic number.

2. Draw graphs of the atomic radius versus atomic number for (*a*) the noble gas atoms, (*b*) the halogen atoms.

3. Would you expect a positive ion to be larger or smaller than the corresponding neutral atom?

4. What is the basis for the assumption that there are no elements yet undiscovered with atomic number less than 69? Less than 1?

5. What experiments can one do to distinguish the metals from the nonmetals?

6. Making use of family relationships, predict the properties of the elements polonium and francium.

7. Potassium has a density of 0.86 grams per $cm.^3$. What is its gram atomic volume?

8. Examine the relative ionization energies of the noble gas atoms in Fig. 10-2. Then predict the relative ionization energies of the halogen atoms.

9. An unspecified element of the nitrogen family (column V) forms a compound with gallium (column III). (*a*) What is the probable empirical formula of that compound? (*b*) If the compound contains 48.1% gallium, what is the atomic weight of the unspecified element?

10. Why are the lanthanides and actinides not included within the main body of the periodic chart?

11. Decide, by analogy with compounds of known formula, and by using the periodic chart, whether or not *stable* compounds with the following formulas are likely to exist.

(*a*) Sb_2O_3	(*b*) BaTe	(*c*) $CsCl_3$
(*d*) $CsCl_2$	(*e*) NeF_2	(*f*) NaO
(*g*) CCl_2	(*h*) ICl	(*i*) GaAs
(*j*) NI_4		

12. Using the periodic chart *only*, predict the formulas of the following compounds: rubidium fluoride, cesium sulfide, barium oxide, phosphorous chloride, boron nitride, aluminum iodide.

13. List two elements other than germanium and silicon that you might expect to be semiconductors.

The exact sciences also start from the assumption that in the end it will always be possible to understand nature, even in every new field of experience, but that we may make no *a priori* assumptions about the meaning of the word "understand."

W. HEISENBERG (1901–)

Eleven

THE STRUCTURE OF ATOMS: III. ATOMS WITH MORE THAN ONE ELECTRON

WE HAVE seen that when the elements are arranged in order of increasing atomic number, a striking harmony becomes apparent. At definite intervals there appear elements with closely related chemical properties. This permits the organization of the many elements into *families* and the construction of a periodic chart.

Whenever we encounter a striking macroscopic simplicity or orderliness like this, we seek some submicroscopic orderliness to account for it. In the present case, it seems reasonable to assume that the periodicity can be explained in terms of some orderliness in the arrangement of the electrons around the nucleus in the atoms of each element.

Review of earlier theories

Let us begin by reviewing the structure of the atoms beyond hydrogen, as deduced from the experiments of Rutherford. The nuclear charge of each element increases by one unit over that of the preceding element, and since each atom is electrically neutral, the number of electrons must also increase by one. Thus, helium has a nuclear charge of two, and two electrons; lithium has a nuclear charge of three, and three electrons; and so forth, right through the periodic chart, to element number 105, which has a nuclear charge of 105, and 105 electrons.

How are these electrons distributed about the nucleus? Evidently we cannot assume that the electrons are traveling helter-skelter, because we need some orderliness to explain the periodicity of chemical properties. It is therefore gratifying to note that Bohr's theory, which has great success in explaining the properties of the hydrogen atom, contains strong elements of orderliness. In particular, the electron can travel around the nucleus only in certain quantized orbits of definite size.

Unfortunately, it was found that Bohr's theory is less successful when extended to more complex atoms. Although the basic idea of quantized energy states was still applicable, the assumption that the electrons travel in planar orbits failed to predict the energy levels for even so simple an atom as helium. Furthermore, it was soon recognized that the planar or flat model envisioned by Bohr is incapable of explaining either the spatial or the bonding properties of the elements. Perhaps even more bothersome to many scientists was the totally artificial way in which the ideas of quantization and the quantum numbers were introduced.

Wave mechanics

It was during an effort to resolve these difficulties that the young French physicist, Louis de Broglie, in 1924, introduced the ideas that led to the next great step forward in our understanding of atomic structure.

De Broglie observed that one of the few places in physics where integers (1, 2, 3, . . . , etc.) occur naturally during the mathematical treatment of the phenomenon is in the physics of waves and vibrations, which deals with such subjects as sound waves or the vibrations of a violin string. It was tempting to extend this notion to the behavior of electrons, by assuming that in certain circumstances electrons behave like waves rather than particles.

The assumption that the electron has wave-like properties was, of course, extremely startling and is still difficult for some of us to assimilate. However, it is somewhat reassuring to learn that this duality of character is shared by ordinary light, which we usually think of in wave terms, but which, strangely enough, sometimes shows particle-like properties. Thus light, on striking the metal plate of a photoelectric cell, ejects electrons from the metal in a way that can be explained only if we assume the light to be in the form of tiny particles or *photons.* Conversely, electrons, which we have hitherto thought of as particles, can be diffracted, that is, have their direction changed by passing through a narrow slit, just as light waves are diffracted.

With the recognition of this dual character of electrons, it became possible to apply the physics of waves (wave mechanics, or quantum mechanics) to the behavior of electrons. As a result, Schrödinger (1927) was able to set forth a mathematical expression, called the wave equation, which in principle describes the behavior of the electrons in even the most complicated of atoms. In practice, the exact solution of this equation for complex atoms is almost impossible. Nevertheless, approximate solutions can be found that tell us a great deal about electronic structure. In the following sections, some of these findings will be discussed.

THE UNCERTAINTY PRINCIPLE

One of the most important consequences of ascribing wave-like properties to electrons is the conclusion that we can no longer locate with unlimited precision the position and velocity of the electron. In terms of our macroscopic notions of physical behavior, this idea is new and strange. In the macroscopic world, which is described by the laws of motion developed by Isaac Newton, we are accustomed to thinking that the exact location and velocity of a material object can be given with any desired degree of accuracy. The accuracy is limited only by the amount of care we are willing to take in making the measurement. Not so in the submicroscopic world. Owing to the tiny size of the electron, the "light" that we must use to "see" the electron has enough energy to cause the electron to move out of its path. Thus, every time we try to locate the electron, the very act of "looking" causes the electron to change its position and/or velocity by indeterminate amounts.

Our inability to specify exactly both the position and the velocity of a particle at a given instant is a general limitation inherent in the nature of measurement, because any conceivable method of measurement will perturb the particle somewhat. The principle expressing this limitation is known as the *uncertainty principle.* However, in practice, the uncertainty resulting from this limitation is appreciable only when the mass of the particle is very, very small (such as that of an electron).

ELECTRON ORBITALS

The uncertainty principle makes it impossible to formulate a model of the submicroscopic world, such as that envisioned by Bohr, in which an electron moves with a precisely known velocity along a precisely predictable path. Instead, we must content ourselves with dealing with probabilities, large or small, of finding an electron in a particular place at a particular time.

The wave equation of Schrödinger provides a way for calculating such probabilities. Mathematical solutions of the wave equation are called *wave functions* or *orbitals.* These mathematical functions can be manipulated in such a way that they yield precisely the probability information we need. The probability information, in turn, can be used to construct pictures of the region in space in which we are likely to find an electron, with any previously specified precision.

Figure 11-1 shows examples of such pictures for the hydrogen atom in various states. The ground-state orbital is denoted by 1*s*; the symbols 2*p* and 3*d* refer to electronically excited states and will be explained in the next section. The boundaries of the three-dimensional figures are calculated so that the probability of finding the electron on the inside is 95% when the hydrogen atom exists in the given state. This implies, of course, a probability of 5% that the electron is somewhere on the outside of the figure. But 5% is small compared to 95%, and the figures therefore represent the regions of space within which the electron exists most of the time. Strictly speaking, figures such as 11-1 should be labeled "95% probability boundaries for the position of the electron," but it has become customary to refer to these shapes as *orbitals,* also. Thus, when we use the term "1*s* orbital of the hydrogen atom," we refer both to a particular mathematical function obtained by solving the wave equation and to a picture of the 95% probability boundary for the position of the electron in that particular state.

Wave mechanics of the hydrogen atom

One of the great achievements of the Bohr theory was to reproduce precisely the observed energy levels of the hydrogen atom. In developing his theory, Bohr assumed that the electrostatic attraction between the positively charged nucleus and the negatively charged electron is given by Coulomb's law. If, similarly, we let Coulomb's law represent the electrostatic interaction in wave mechanics, solution of the wave equation for the permitted values of the energy of the hydrogen atom leads to exactly the same formula and exactly the same values as does Bohr's theory. There are, however, a number of important additional features: Instead of obtaining a single quantum number, we find that, in fact, three quantum numbers are required to describe completely the state of the atom. These quantum numbers are usually denoted by n, l, and m. Their physical significance may be described as follows:

The quantum number n is called the *principal quantum number*. It is analogous to the quantum number that appears in Bohr's theory in that it determines the energy of the hydrogen atom. Thus, *for the hydrogen atom,* the energy is given by

$$E = \frac{-2\pi^2 me^4}{h^2 n^2}$$

which is identical with the expression obtained by Bohr (p. 150). As before, m is the mass and e is the charge of the electron, and h is Planck's Constant. Values of n are restricted to the positive integers, that is, $n = 1, 2, 3, \ldots$†

Besides determining the energy of the hydrogen atom, the quantum number n also serves as a crude index of the average distance between the electron and the nucleus, provided that the average is taken over an extended period of time. Roughly speaking, the average distance increases with n. However, when speaking about average distances, we must consider also the shape of the orbital occupied by the electron, and to do that we need the quantum number l.

For all values of n except $n = 1$, the wave equation for the hydrogen atom has more than one solution. These orbitals of equal n are

† In a still more rigorous solution of the wave equation, the energy is found to depend also very slightly on the value of the quantum number l. The corresponding separation of the energy levels has actually been observed, but it is so small that the measuring technique must be highly refined. We shall neglect this effect.

different, either because of differences in shape, or because they extend in different directions in space. For a given n, the *shape* of an orbital is determined by the quantum number l, and the *direction in space*, by the quantum number m. It turns out that l can have the values 0, 1, and so on, up to $n - 1$. For example, for $n = 3$, $n - 1 = 2$, and therefore l can be 0, 1, or 2. For $n = 2$, $n - 1 = 1$. Therefore l can be 0 or 1. For $n = 1$, $n - 1$ is zero, and the only value l can take is zero.

The values that the quantum number m can take are limited by the value of l. They are $-l, -(l - 1), \ldots, -1, 0, +1, \ldots, +(l - 1), +l$. Thus, if l is zero, m can only take the value zero. If $l = 1$, m can be -1, 0, or $+1$. If $l = 2$, m can be $-2, -1, 0, +1, +2$, and so on.

These relationships are summarized in the table on the next page.

Fig. 11-1
Probability density shapes for the hydrogen atom [after E. Cartmell and G. W. A. Fowles, *Valency and Molecular Structure* (New York: Academic Press, 1956].

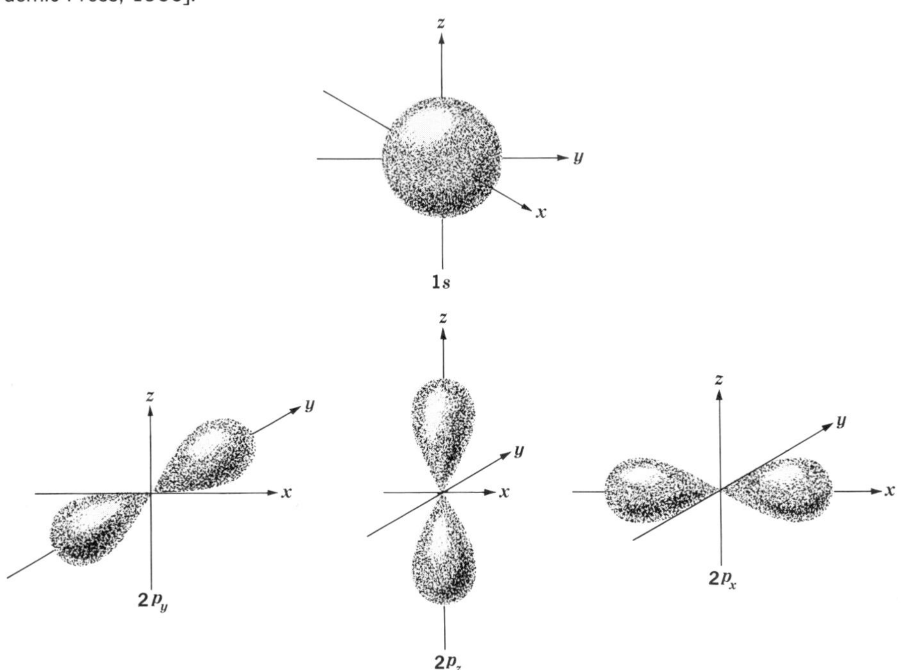

RELATION OF THE QUANTUM NUMBERS

n	1	2		3			4
l	0	0	1	0	1	2	
m	0	0	−1 0+1	0	−1 0+1	−2−1 0+1+2	
Number of orbitals for the given n	1	4		9			16

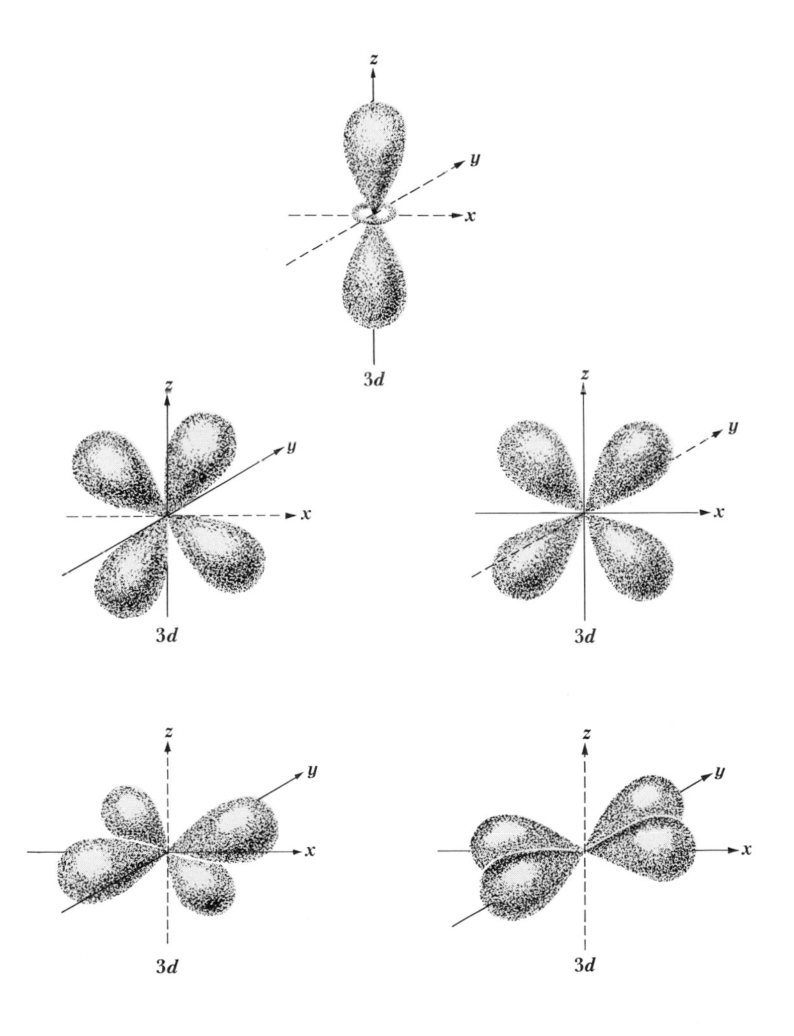

Note that when $n = 1$, there is one orbital; when $n = 2$, there are four orbitals; when $n = 3$, there are nine; and when $n = 4$, there are sixteen. In general, the total number of orbitals associated with a given principal quantum number n is equal to n^2.

SHAPES AND DESIGNATIONS OF ORBITALS

Some orbital shapes for the hydrogen atom have been shown in Fig. 11-1. Recall that the shapes shown in this figure depict the 95% probability boundaries for the position of the electron. The shapes depend on the value of l, as follows:

When $l = 0$, the orbital shape is a sphere, that is, the orbital has identical properties in all directions. An example is the shape labeled 1*s* in Fig. 11-1, which represents the orbital that is obtained for $n = 1$. For higher values of n, orbitals with $l = 0$ likewise have spherical symmetry but differ in how the probability density varies with distance from the nucleus. The radii of the spheres increase with n.

When $l = 1$, the orbital shape consists of two lobes oriented along an axis through the centers of the lobes and the nucleus of the atom. An example is the shape labeled 2*p* in Fig. 11-1, which is calculated for $n = 2$ and $l = 1$. This shape is shown as pointing in three different directions in space if there is an external magnetic field present: these directions depend on the value of the quantum number m, which can be either -1, 0, or $+1$. One of the shapes can be thought of as lying along the *x*-axis, the second is perpendicular to it along the *y*-axis, and the third is perpendicular again along the *z*-axis. Orbitals for $l = 1$ and n greater than 2 have similar shapes consisting of two collinear lobes, but their boundaries are farther from the nucleus. For any given value of n, the directions of the three orbitals corresponding to $l = 1$ and $m = -1$, 0, and $+1$ are mutually perpendicular.

When $l = 2$, the orbitals are even more complex in shape, and there are five of them, since m can take the values -2, -1, 0, $+1$, and $+2$. The five orbitals corresponding to $n = 3$ and $l = 2$ are shown in Fig. 11-1 over the caption, 3*d*.

Historically, the orbitals having various values of l were designated by means of letters. Thus, the orbital corresponding to $l = 0$ is called an *s* orbital. For $l = 1$, the orbitals are called *p* orbitals. Those orbitals for which $l = 2$ are known as *d* orbitals, and the next higher ones are called *f*, *g*, etc. In this system of notation, an electron whose principal quantum number is 1 would be designated as 1*s*. If $n = 2$, the electron would be either 2*s* or one of the 2*p*'s, depending upon the values of l and m. The notation is summarized in the following table:

NOTATION FOR ELECTRON ORBITALS

n	1			2				3		
l	0	0		1		0		1		2
m	0	0	−1	0	+1	0	−1	0	+1	−2 to +2
Orbital designation	1*s*	2*s*	2*p*	2*p*	2*p*	3*s*	3*p*	3*p*	3*p*	3*d* (five orbitals)

It is customary to use the term *shell of orbitals,* or just *shell,* to denote the entire set of orbitals associated with a given value of n. Therefore, when we say that the electron is in the first shell, we mean that its principal quantum number is 1. When we say that the electron is in the second shell, we mean that its principal quantum number is 2, and so on. Referring to the preceding table, we see for example that an electron in the second shell can be in the 2*s* or any of the 2*p* orbitals.

In the case of the hydrogen atom, all orbitals belonging to the same shell have the same energy. For more complex atoms, however, this simple situation no longer exists, and the *s, p, d,* etc. orbitals within the same shell differ in energy. It thus becomes desirable to show pictorially the energy relationships among the various orbitals. This can be done by letting a box represent each orbital in which an electron can reside. The values of l are then indicated from left to right horizontally along the base of the diagram, while the values of n are plotted vertically, with energy increasing upward. A dot is used to represent the electron.

Using this notation, we depict the ground state of the hydrogen atom as shown in Fig. 11-2. The electron is represented by the dot in the box that represents the 1*s* orbital. An excited state of the hydrogen atom in which the electron is in the 2*s* orbital can then be represented as shown in Fig. 11-3.

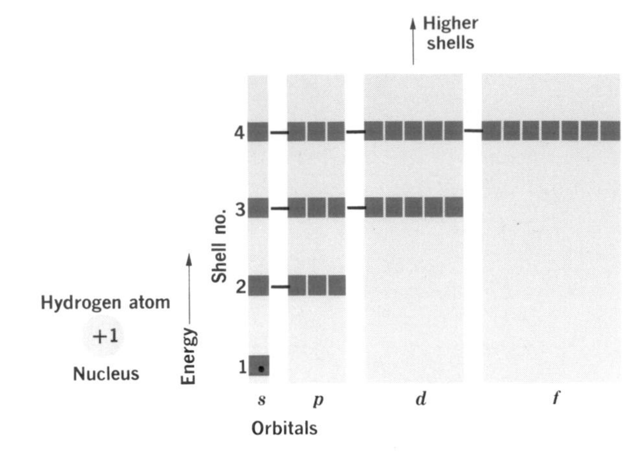

Fig. 11-2
Orbital diagram of the hydrogen atom in the ground state.

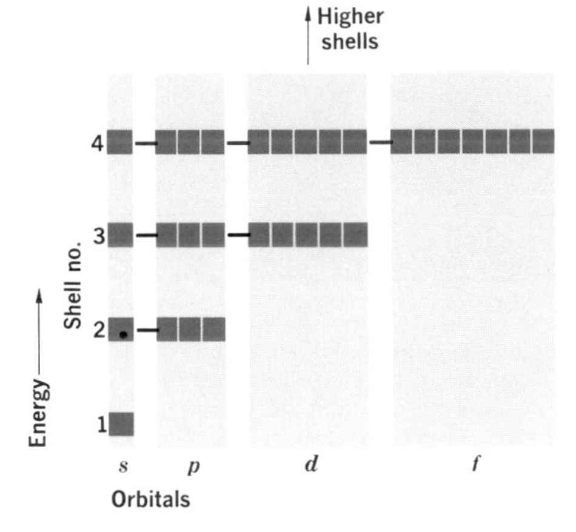

Fig. 11-3
Orbital diagram for an excited hydrogen atom.

ELECTRON SPIN

At about the same time (1925) that the detailed picture of the orbitals of the hydrogen atom was worked out, two Dutch physicists, G. E. Uhlenbeck and S. Goudsmit, suggested that the finest details of the properties of the hydrogen atom, particularly its behavior in a magnetic field, could be explained only if it was further assumed that the electron is spinning on its axis. According to their theory, all electrons are spinning at the same rate. However, the spin may be either in a clockwise direction or in a counterclockwise direction.

Atoms more complex than the hydrogen atom

ELECTRON ORBITALS FOR THE LIGHTER ATOMS

In principle, the wave equation can be used to find the electron orbitals and values of the energy for any atom, regardless of its complexity. However, in practice, the difficulty of the required mathematics increases rapidly with the number of electrons in the atom. For the neutral helium atom (which has two electrons), the mathematics is already extremely difficult, and for atoms with more than two electrons, the difficulties require the use of high-speed computers. However, it can be shown that, for the lighter atoms, the orbitals are much like those of the hydrogen atom. When there are two or more electrons in the atom, there are some changes because of the mutual repulsion of the negatively charged electrons. But the basic model, which involves the idea of shells of orbitals and the quantum numbers n, l, and m, is still correct.

In atoms containing more than one electron, the energy associated with a given orbital depends not only on the principal quantum number, n, but also on the quantum number l. It is found that, for a given value of n, the *s* orbital has a lower energy (is more stable) than the *p* orbitals, which in turn are more stable than the *d* orbitals, and so on. For the first few elements of the periodic table, these differences are moderately small, and the orbitals available to the electrons can be represented qualitatively by the diagram in Fig. 11-4. The reader will note the close resemblance to the orbital diagram for the hydrogen atom.

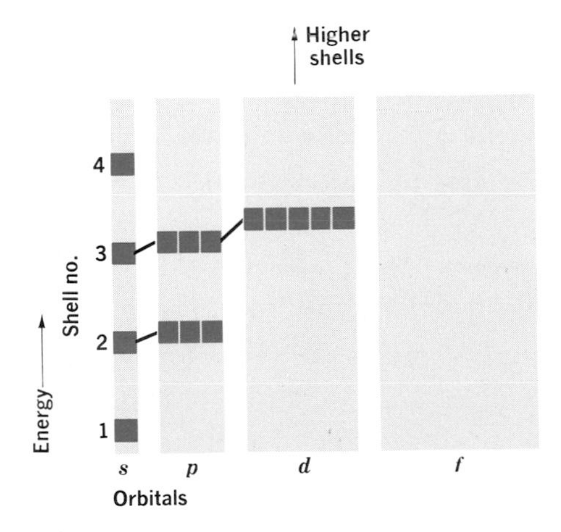

Fig. 11-4

Generalized orbital diagram.

THE PAULI EXCLUSION PRINCIPLE

Assuming the preceding picture of the orbitals for the lighter elements, the question now arises: How do the electrons distribute themselves among these orbitals? In this section we shall be concerned with the ground state of the atom, that is, the state of lowest energy. If we imagine that we start out with the bare nucleus and add electrons one at a time, until we have the neutral atom, each electron will go into the orbital of lowest energy in which there is still room for that electron. To be able to describe the ground state of an atom, we must therefore ask: How many electrons can each orbital accommodate?

According to the Austrian physicist, Wolfgang Pauli, the answer is quite simple:

There can be no more than two electrons in any one orbital, and these electrons must be spinning on their axes in opposite directions.

The statement that each orbital can contain only two electrons spinning in opposite directions is known as the *Pauli exclusion principle*. Thus we see that the first shell, which consists of only one orbital, can accommodate, at most, two electrons, which must differ in the direction of their spins. The second shell can hold as many as eight electrons: two in the $2s$ orbital and six in the three $2p$ orbitals. The third shell can hold up to eighteen electrons: two in the $3s$ orbital, six in the three $2p$ orbitals, and ten in the five $3d$ orbitals. With this concept, it now becomes possible to describe the electron configurations of the lighter atoms.

ELECTRON CONFIGURATIONS OF THE LIGHTER ATOMS IN THEIR GROUND STATES

Electron configurations for the elements helium, lithium, beryllium, boron, and carbon are depicted in Fig. 11-5. According to the exclusion principle, both electrons of the helium atom can be found in the $1s$ orbital. Of the three electrons of the lithium atom, two are in the $1s$ orbital, and the third is in the $2s$ orbital, which is the next vacant orbital with the lowest energy. For the beryllium atom we find a configuration resembling that of the lithium atom, except that there are two electrons in the $2s$ orbital. For the boron atom, which has five electrons, we find that two electrons are in the $1s$ orbital, two in the $2s$ orbital, and the fifth is in a $2p$ orbital. When we come to the element carbon, which has six electrons, an ambiguity arises that merits consideration. Since the $2p$ orbitals are all represented as having the same energy, the sixth electron can go either into the already half-filled $2p$ orbital or into one of the other $2p$ orbitals that is still vacant. Analysis of the spectrum of the carbon atom shows that the two electrons actually occupy different $2p$ orbitals, as shown in Fig. 11-5. This result is typical of most atoms: when electrons are added one at a time to a p subshell, or to a d subshell, or to an f subshell, the first few electrons will go into separate orbitals, until all the orbitals of the subshell contain one electron each. Orbitals will be occupied by two electrons only if there are no more vacancies in the subshell.

SHORTHAND NOTATION FOR ELECTRON CONFIGURATIONS

A convenient and more condensed way to represent electron configurations than that used in Fig. 11-5 is as follows: Write down the designation of all orbitals in which there are electrons; show the number of electrons in a given orbital as a superscript following the orbital designation.

For example, the electron configuration of the hydrogen atom in its ground state is denoted by $1s^1$—one electron in the $1s$ orbital. Note that vacant orbitals are not mentioned in this notation. The electron configurations of some other atoms are represented as follows:

Helium $1s^2$
Lithium $1s^2 2s^1$
Beryllium $1s^2 2s^2$
Boron $1s^2 2s^2 2p^1$
Carbon $1s^2 2s^2 2p^2$
Nitrogen $1s^2 2s^2 2p^3$
Oxygen $1s^2 2s^2 2p^4$

Note that, in this notation, the orbitals are written down in the order of increasing energy.

ELECTRON CONFIGURATION OF THE HEAVIER ATOMS

As the number of electrons becomes greater, the differences in the energy of the s, p, d, and f orbitals in the same shell become greater also. Thus, for example, the difference in the energy between the 2*s* and the 2*p* orbitals is greater in lithium than in helium, as shown in Fig. 11-5. These differences become so great that by the time we come to the 19-electron atom, potassium, the energy of the 4*s* orbital is lower than that of the 3*d* orbitals. This is just the reverse of what might have been expected on the basis of the simple hydrogen-like shell model. The energy relationships for the potassium atom are depicted in Fig. 11-6.

Energy relationships of the electron orbitals, such as that shown in Fig. 11-6, can be derived experimentally for any atom from an analysis of the atomic spectrum. The methods are intricate and highly specialized, and they form the substance of an entire field of science, known as *atomic spectroscopy*. Detailed discussions of these methods require considerable background knowledge in mathematics, physics, and chem-

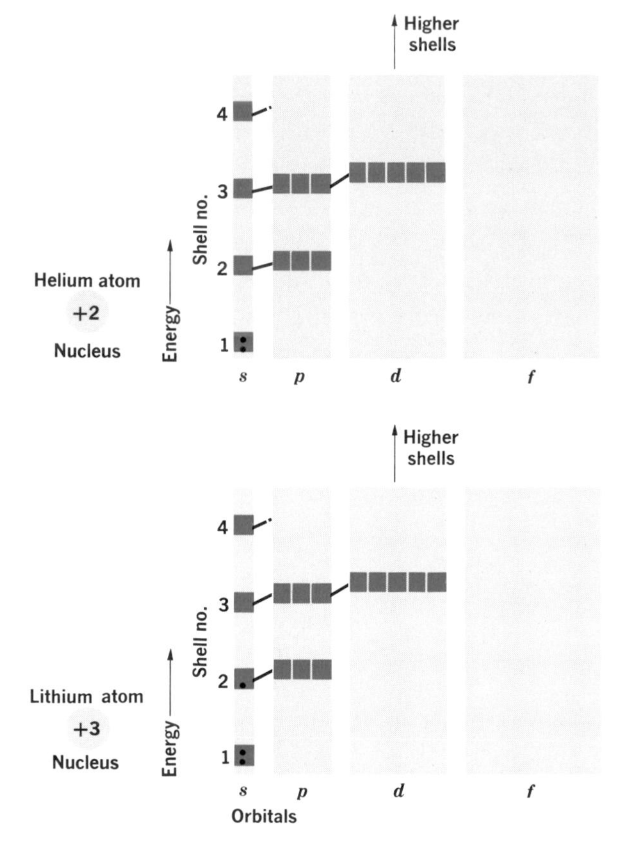

Fig. 11-5
Some orbital relationships. Note the increasing separation of the *s,p,d, . . . ,* orbitals in any given shell with increasing number of electrons. For convenience, the energy spacing of the shells has been depicted as uniform.

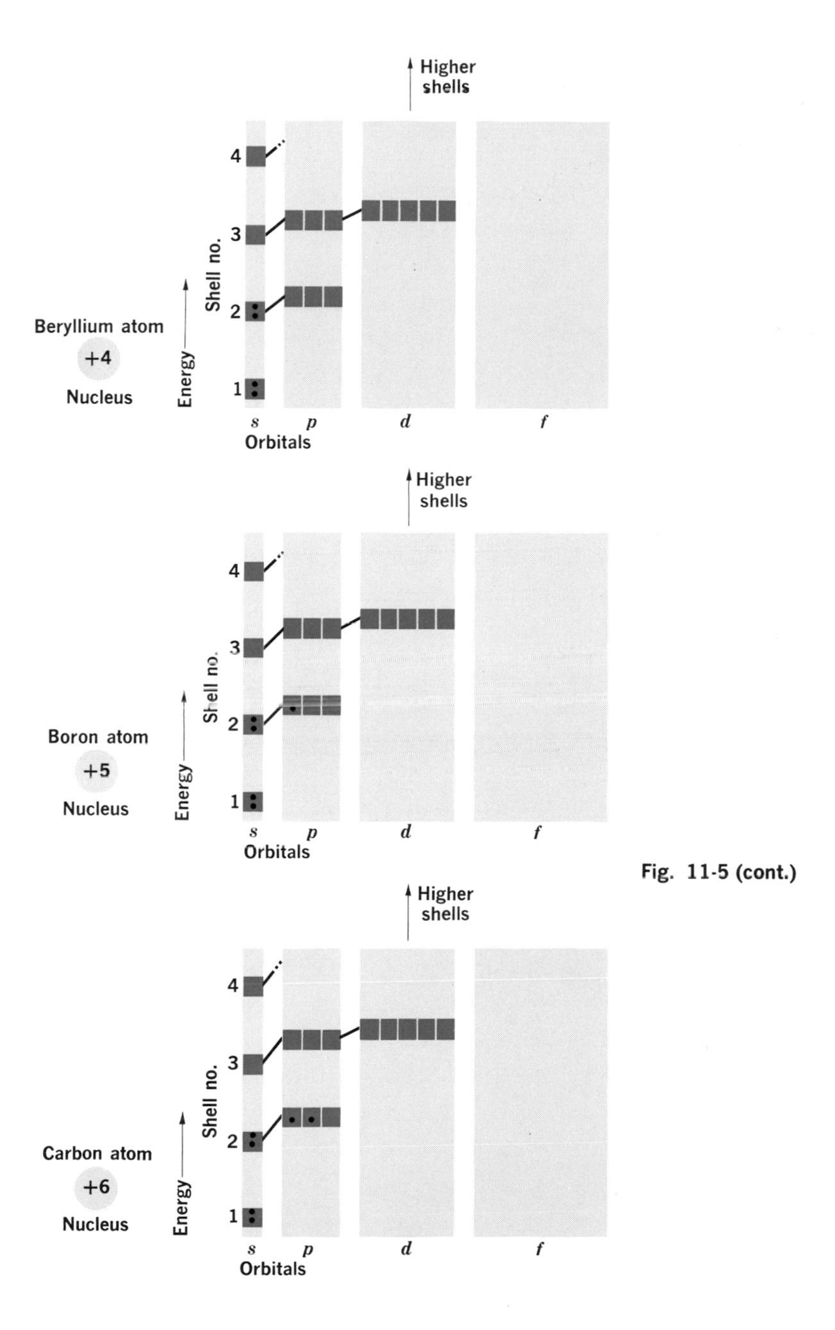
Higher shells
4
3
2
1
Shell no.
Energy
Beryllium atom
+4
Nucleus
s p d f
Orbitals
Higher shells
4
3
2
1
Shell no.
Energy
Boron atom
+5
Nucleus
s p d f
Orbitals
Higher shells
4
3
2
1
Shell no.
Energy
Carbon atom
+6
Nucleus
s p d f
Orbitals

Fig. 11-5 (cont.)

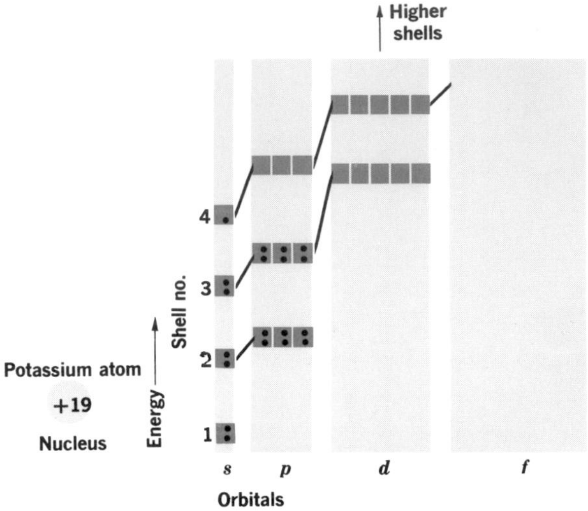

Fig. 11-6
Schematic representation of orbitals in the postassium atom.

istry. We shall therefore content ourselves with merely stating one of the important results.

For atoms other than the first few in the periodic chart, the energy relationship of the electron orbitals can be derived as follows. First, we write down an array of hydrogen-like electron orbitals:

1*s*				
2*s*	2*p*			
3*s*	3*p*	3*d*		
4*s*	4*p*	4*d*	4*f*	
5*s*	5*p*	5*d*	5*f*	5*g*
6*s*	6*p*	6*d*	etc.	
7*s*	7*p*	etc.		
8*s*	etc.			

Next, we draw a set of parallel diagonal lines, each line terminating in an *s* orbital:

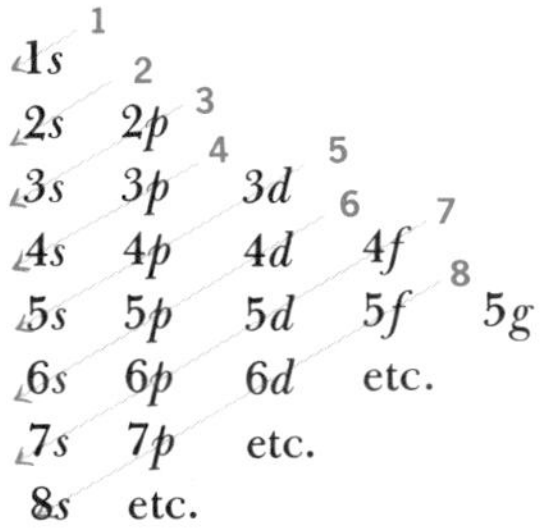

We number these diagonal lines 1, 2, 3, . . . , beginning at the top of the diagram. We can then predict the relative energy of the orbitals in a heavier atom by following along the successive diagonal lines and reading off the orbitals in the order in which they appear along each line. The reader will convince himself that the order is then as follows:

1*s* 2*s* 2*p* 3*s* 3*p* 4*s* 3*d* 4*p* 5*s* 4*d* 5*p* 6*s* 4*f* 5*d* 6*p* 7*s* etc.

The preceding sequence of electron orbitals contains the beginnings of a rational theoretical explanation of the periodic chart. Suppose we rewrite the sequence, drawing a vertical line before every *s* orbital.

SECTIONING OF ELECTRON ORBITALS

Orbital designation	1s	2s 2p	3s 3p	4s 3d 4p	5s etc.
Ordinal number of subdivision	1	2	3	4	5
Maximum number of electrons in subdivision	2	8	8	18	18

This procedure divides the sequence into a number of subdivisions. The first subdivision consists of the 1*s* orbital and can accommodate two electrons. The second subdivision consists of one 2*s* and three 2*p* orbitals and can accommodate eight electrons. The third subdivision consists of one 3*s* and three 3*p* orbitals and can accommodate eight electrons. The fourth subdivision consist of one 4*s*, five 3*d*, and three 4*p* orbitals and can accommodate eighteen electrons.

On comparing the numbers of electrons that can be accommodated in successive subdivisions with the numbers of elements in successive rows of the periodic chart, we find exact equality. In the periodic arrangement of the elements according to Fig. 10-1, the first line consists of two elements, hydrogen and helium. The second line consists of eight elements, the third consists of eight, and the fourth consists of eighteen. This equality suggests that the chemical properties of the elements bear a direct relationship to the electron configurations of the atoms. The nature of this relationship will be explored in the next section.

Electronic structure and the periodic chart

We are now ready to give an explanation of the periodicity in chemical properties of the elements in terms of electronic structure. For this purpose, it is convenient to divide the elements into two classes.

1. THE MAIN-GROUP ELEMENTS. These are the elements of the families that begin with H, Be, B, C, N, O, F, and He, respectively. The main-group elements are listed in columns I–VIII of the periodic chart on page 171.
2. THE SUBGROUP ELEMENTS. These are the transition metals, the rare earths, and the actinides.

THE MAIN-GROUP ELEMENTS

ALKALI METALS It has been noted that lithium has two electrons in the first shell and one electron in the second shell. For the next alkali metal, sodium, the electron distribution is as shown in Fig. 11-7. We see that in sodium we again have a structure with just one electron in the outermost shell, although this time it is the third shell.

On examining the electronic structures of the other alkali metals, we find that in every case there is one electron in the outermost shell.

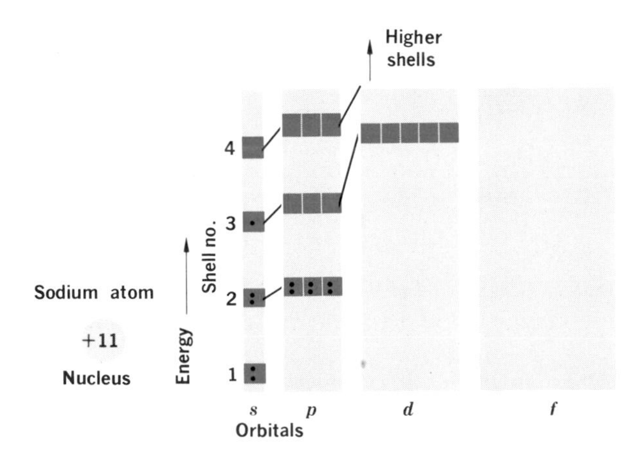

Fig. 11-7
Schematic representation of orbitals in the sodium atom.

Potassium, for example, has one electron in the fourth shell, which is now the outermost shell, as shown in Fig. 11-6.

The fact that each alkali metal atom has one electron in the outermost shell is very striking indeed. Here is the underlying orderliness that can account for the similarity in chemical properties; here is a group of elements, all belonging to the same family, with electronic structures closely related to one another. The observation suggests a possible generalization:

The number of electrons in the outer shell of orbitals is the same for all members of a given family.

ALKALINE EARTH METALS Let us see whether this generalization applies to the members of the alkaline earth family as well. The electronic structures of elements 4, 12, and 20 are depicted in Fig. 11-8. In this case we find two electrons in the outermost shell of orbitals. Evidently the generalization again applies.

A similar situation exists for all of the main-group families discussed in the preceding chapter. All members of a given family have the same number of electrons in the outermost shell of orbitals. Furthermore, this number of electrons is equal to the group number of the family, that is, the Roman numeral at the head of the column in the periodic chart. Thus, group I consists of the alkali metals, all of which have one electron in the outer shell. The alkaline earths constitute group II, and these elements have two electrons in the outer shell. The halogens are members of group VII, and we therefore expect them to have seven electrons in their outermost shell. That is indeed the case.

THE SUBGROUP ELEMENTS

The series of metals beginning with scandium (21) and ending with zinc (30) belong to the family of transition metals and are typical of that family. The element immediately preceding scandium in the periodic chart is calcium, whose electronic structure is shown in Fig. 11-8. Here the outermost electrons are in the 4*s* orbital rather than in the still vacant 3*d* orbitals. In the series beginning with scandium, the 3*d* orbitals fill up.

It is the filling of these inner orbitals that leads to the special chem-

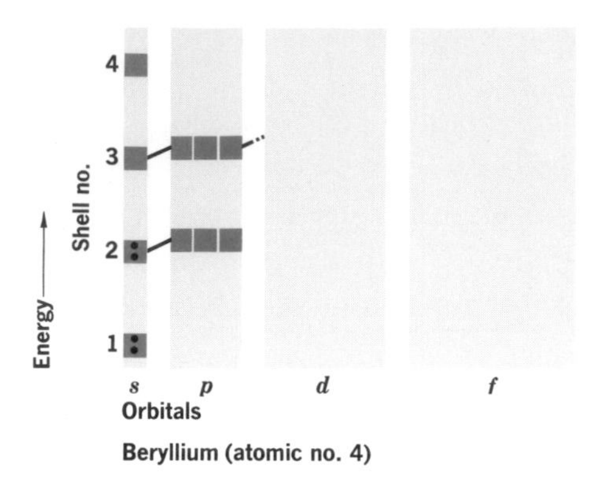

Beryllium (atomic no. 4)

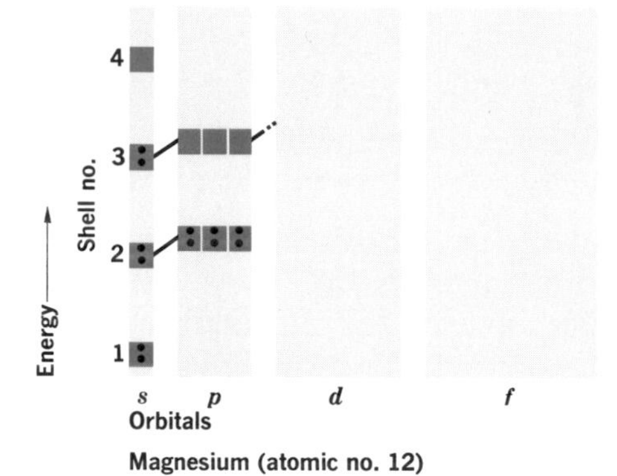

Magnesium (atomic no. 12)

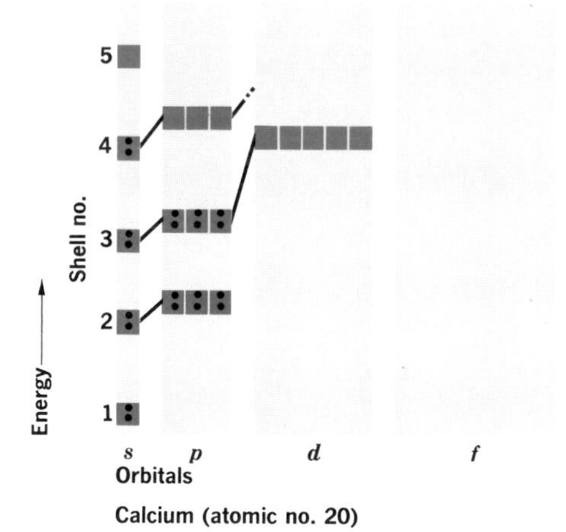

Calcium (atomic no. 20)

Fig. 11-8

Schematic representation of the orbitals in atoms of the alkaline earth metals.

ical properties of the subgroup elements. Thus, scandium has one electron in the 3*d* orbitals, titanium two, vanadium three, and so on, until a total of ten electrons has been added, filling the 3*d* orbitals and, consequently, the third shell. This latter condition occurs with zinc, which has a total of 30 electrons. Examples of the electronic structures are shown in Fig. 11-9.

The element immediately following zinc is gallium (31). In this element, the thirty-first electron goes into the vacant 4*p* orbital, leading to the electronic structure shown in Fig. 11-9. The structure of gallium

Fig. 11-9
Schematic representation of the orbitals in atoms of some transition elements, and of gallium.

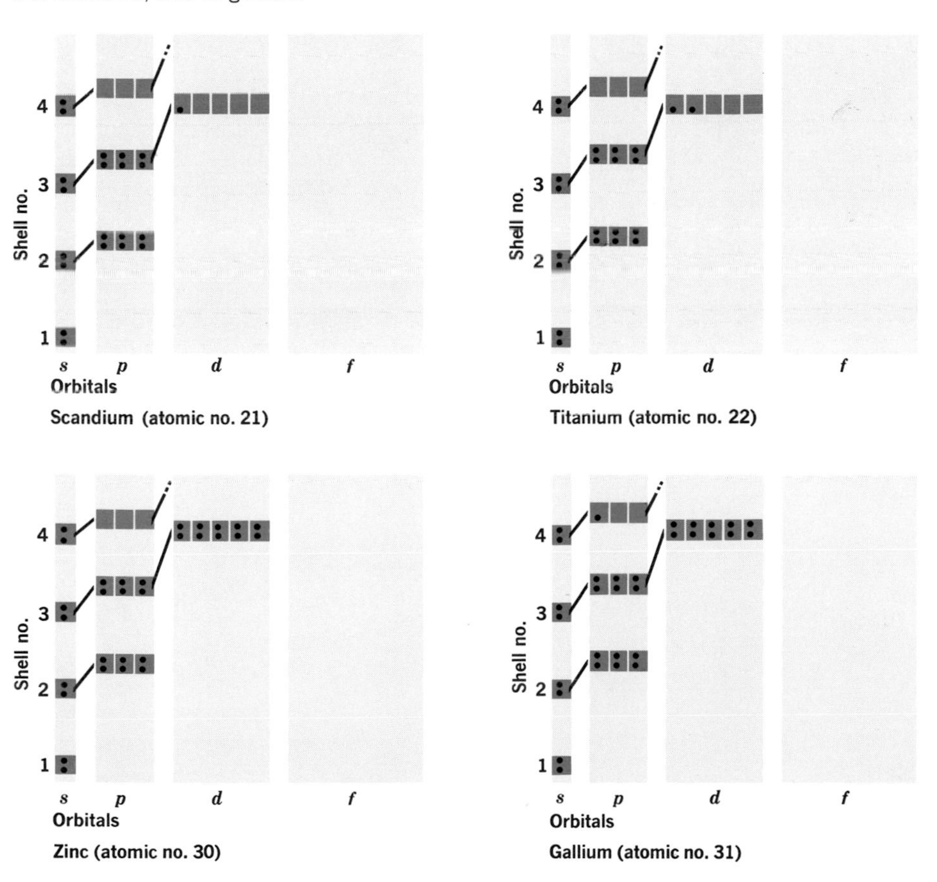

is therefore like that of aluminum, with three electrons in the outermost shell of orbitals. Gallium thus belongs to group III of the main-group elements.

What has been illustrated here is true for all subgroup elements. Each element differs from the one preceding it in the periodic chart by one electron. For the subgroup elements, this electron occupies an empty orbital of an inner shell rather than an orbital of the outermost shell.

Fig. 11-10
Schematic representation of the orbitals in some noble gas atoms.

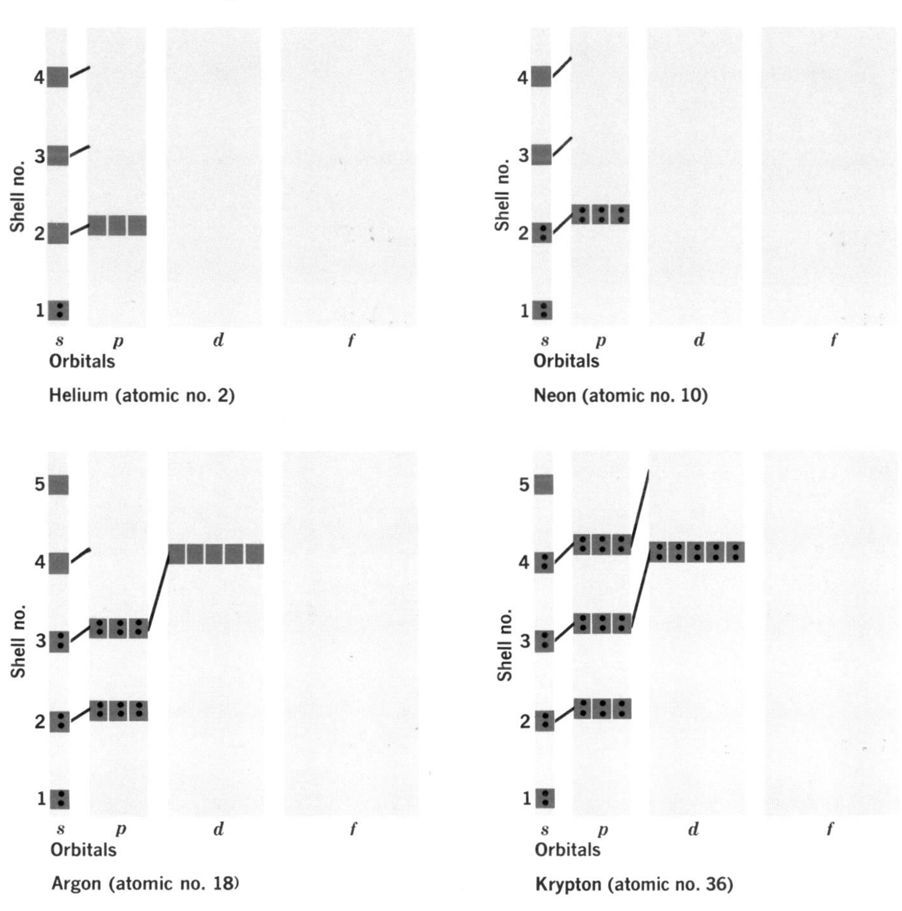

Gilbert Newton Lewis: 1875–1946
Among Lewis' many contributions are the rule of the octet and theories of ionic and covalent bonding. [Photograph courtesy University of California.]

THE OCTET RULE

The family of elements called the noble gases deserves special consideration. The electronic structures of the first four are shown in Fig. 11-10. In every case, with the exception of helium, the outer shell of orbitals contains eight electrons. There must be something peculiarly stable about an octet of electrons in the outermost shell, since we have seen that the noble gases are much less reactive than other elements. Helium is an exception to the "rule of eight." However, the first shell cannot hold more than two electrons; and evidently filling the first shell also leads to a stable condition.

In 1916, the American chemist, G. N. Lewis, suggested that atoms other than those of the noble gases should similarly become more stable if they could somehow acquire eight electrons in the outermost shell of orbitals. We shall see in the next chapters that this is a very powerful concept for understanding chemical reactivity and compound formation.

Suggestions for further reading

Gamow, G., "The Principle of Uncertainty," *Scientific American,* Jan. 1958, p. 51.

George, D. V., "Teaching the Uncertainty Principle and Natural Philosophy," *Journal of Chemical Education, 46,* Oct. 1969, p. 633.

Luder, W. F., "Electron Repulsion Theory: Atomic Orbitals and Atomic Models," *Chemistry, 42,* June 1967, p. 16.

Müller, E. W., "Atoms Visualized," *Scientific American,* June 1957, p. 113.

Hochstrasser, R., *The Behavior of Electrons in Atoms* (New York: W. A. Benjamin Co., Inc., 1964).

Hoffmann, B., *The Strange Story of the Quantum* (New York: Harper & Row, Publishers, 1947).

Winsor, F., *The Space Child's Mother Goose* (New York: Simon and Schuster, Inc., 1958).

Questions

1. What changes in Bohr's theory of the hydrogen atom, are necessitated by the adoption of the uncertainty principle?

2. Define the term *orbital* in two different ways.

3. Briefly describe the distinguishing spatial features of s, p and d orbitals.

4. What are the permitted values of the l and m quantum numbers when the principal quantum number, n, is equal to 4?

5. What is the smallest possible value for n if $m = 3$?

6. In general terms, what property of the electron is described by each of the quantum numbers n, l, and m?

7. Write the electron configuration of elements 3 through 18 (lithium to argon); then compare the electron configurations of those elements that belong to the same family. Do you find any basic similarities that might account for the common chemical properties in each family?

8. Using diagrams similar to Fig. 11-4, depict the electron configuration of (*a*) Ca, (*b*) Ga, (*c*) Br, (*d*) Kr, and (*e*) Tc.

9. Using diagrams similar to Fig. 11-4, depict the electron configuration of three elements of the oxygen family (column VI).

10. Compare the energy of *s* and *p* orbitals in the following atoms: (*a*) Hydrogen (*b*) Sodium.

11. *Without* reference to the periodic table, assign the following elements to the appropriate family: 16, 20, 35, 37, 50.

12. Carefully explain the relationship between chemical periodicity and electronic structure.

13. State the Pauli exclusion principle. Does the exclusion principle necessarily imply a periodicity of chemical properties?

14.* (*a*) Draw a diagram of a standing wave and indicate the dimension which corresponds to the wavelength. (*b*) How is the "wavelength" of the electron related to its energy?

15.* (*a*) What is the evidence that light shares the dual properties of waves and particles? (*b*) Why doesn't a baseball show "uncertainty" in its position?

* Questions marked with an asterisk (*) are more difficult or require supplementary reading.

"For the sake of persons of different types of mind scientific truth should be presented in different forms and should be regarded as equally scientific whether it appears in the robust form and vivid colouring of a physical illustration or in the tenuity and paleness of a symbolic expression."

JAMES CLERK MAXWELL (1831–1879)

Twelve

PRINCIPLES OF COMPOUND FORMATION: I. IONIC COMPOUNDS

HOW DO atoms combine to form molecules? Now that we have studied the electron orbitals in the atom, we are ready to consider this question. In this and the next chapter we shall study the nature of chemical bonds. *We shall limit our considerations to the eight main families of elements,* which are listed in columns I–VIII of the periodic chart on page 171. However, similar concepts of interatomic bonding apply also to the transition metals, the rare-earth metals, and the actinide metals.

Two basic assumptions for main group elements

We have seen that all members of the same family of elements have an identical number of electrons in the outermost shell. This number is the one conspicuous feature of electronic structure that is uniquely characteristic of each family. Since chemical properties are also characteristic of each family, it seems plausible to assume a close relationship between chemical properties and outer electrons.

This argument becomes even more convincing when we study the electrons that are *not* in the outermost shell. The distribution of these inner electrons among their shells for main group elements is identical to that of a noble gas. We may assume, therefore, that the inner electrons are quite unreactive and that the chemical reactivity and combining power of an atom is determined by the electrons in the outermost shell. In summary:

1. The electrons in the outermost shell are the only ones that are directly involved in compound formation.
2. Stable molecules are formed only if the electron configuration is more stable in the molecule than in the separated atoms. The most stable electron configurations are like those of the noble gases.

We shall see that there are two processes by which each atom in the molecule can acquire a stable noble gas electron configuration: electron transfer, and electron sharing. Both processes will be discussed in detail. In this chapter we shall consider electron transfer. In the following chapter we shall consider compound formation by electron sharing.

The formation of ions by electron transfer

Let us first consider the lithium atom (atomic number 3), which has the electron configuration shown below:

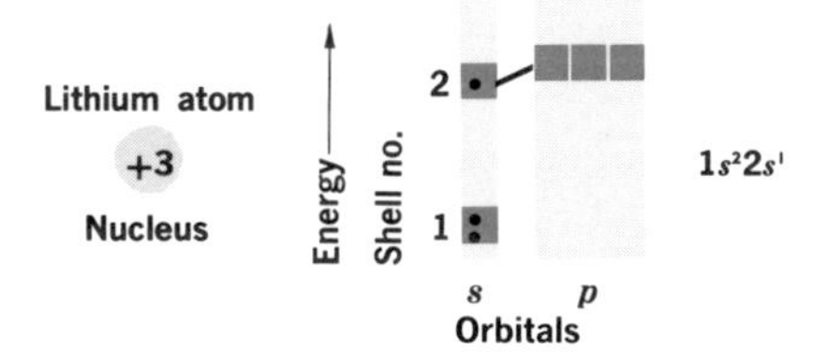

$1s^2 2s^1$

Shells higher than the second are empty and are not shown. Note that the first shell is filled, whereas the second shell contains only one electron. Now let us compare this electron configuration with that of an atom of helium (atomic number 2), the nearest noble gas:

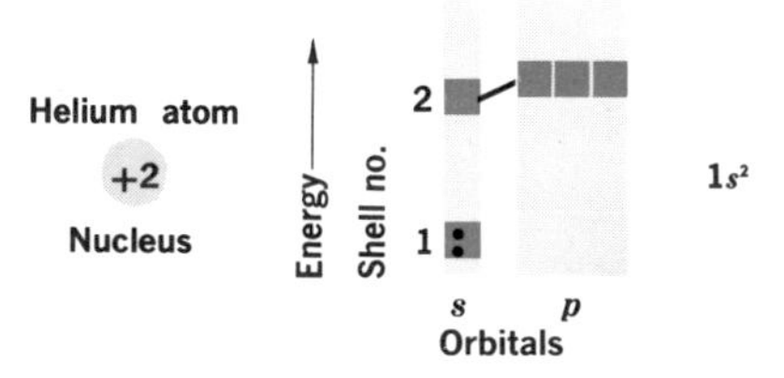

$1s^2$

The two configurations differ only by the additional electron in the second shell of the lithium atom. The loss of this electron by the lithium atom would result in the formation of a lithium ion, which has an electron configuration identical to that of the inert and stable helium atom:

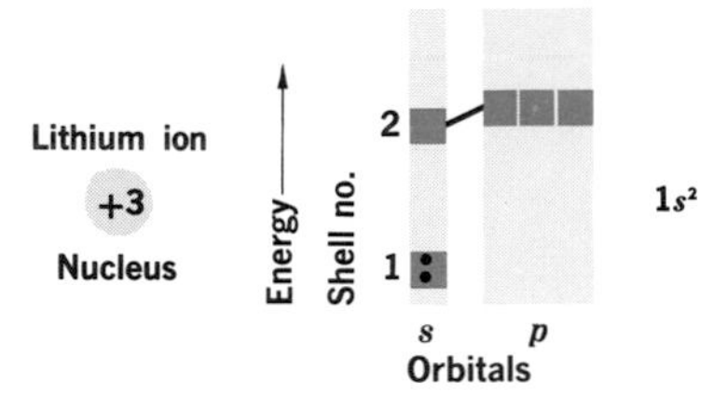

Of course, the lithium ion will have a net charge of +1 because now there are only two electrons associated with the lithium nucleus, which has a charge of +3. We would expect that the lithium ion is stabilized by the same forces that are responsible for the stability of the helium atom.

Next let us consider the fluorine atom. Here the electron configuration is as follows:

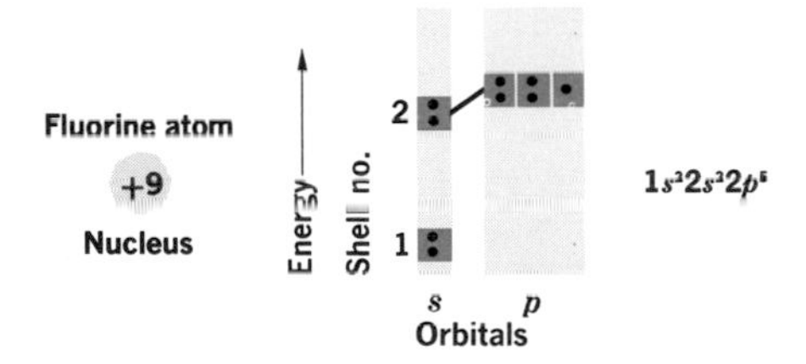

The shells higher than the second are empty and again are not shown. To see how the fluorine atom might acquire a more stable electron configuration, let us consider the noble gas nearest to fluorine, neon:

Neon atom
+10
Nucleus
Energy
Shell no.
2
1
s
p
Orbitals
$1s^22s^22p^6$

The neon atom has eight electrons in the second shell, whereas the fluorine atom has only seven. The fluorine atom should therefore be stabilized by the gain of an electron to complete the octet in the second shell, and a fluoride† ion, with a net charge of -1, is formed.

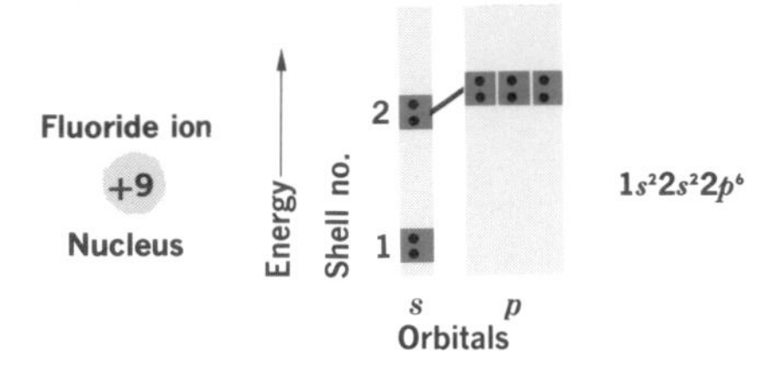

Next, let us consider what might happen when a lithium atom comes into contact with a fluorine atom. If the lithium atom transfers its outer electron to the fluorine atom, as shown below, the two ions that result will both have a stable noble gas electron configuration. Since the electric charges of the ions are of opposite sign, the ions will also attract each other and form a molecule. If we bring into contact macroscopic samples of lithium and fluorine, the same process of electron transfer can be repeated many times, and the ionic compound, lithium fluoride, is produced.

$$\text{Li} + \text{F} \longrightarrow \text{Li}^{\oplus} + \text{F}^{\ominus}$$

Since each lithium atom *loses* one electron and each fluorine atom *acquires* one electron, we anticipate that lithium fluoride will have the formula LiF, that is, for each lithium ion there will be one fluoride ion. This prediction is fully in agreement with experiment.

The preceding ideas may be generalized as follows:

There is a transfer of electrons from the atoms of one element to those of the other to produce ions with electron configurations like those of noble gases; the ions are then held together by the electrostatic attraction of their opposite charges.

It should be emphasized that electron transfer can take place spontaneously only if an ionic compound is formed: the formation of the separated ions requires a net input of energy and does not take place spontaneously.

† When an atom gains electrons to become a negative ion, the ending of the name is changed to *-ide*.

We will next apply the theory of ionic compound formation to cases other than lithium fluoride, but first let us simplify our rather cumbersome method of representing electron configurations.

Symbolic representation of atoms and ions

Since we seek a simple representation, we shall depict only those characteristics that are useful in the discussion of chemical bonding.

We saw in Ch. 11 that the chemical properties of the main-group elements depend on the number of electrons in the outermost shell. For the elements in columns I through VIII of the periodic chart (page 171), that number ranges from one to eight, and the electrons occupy as many as four orbitals. Thus, we can summarize the essential electronic feature of the elements by writing the symbol of the element, surrounded by four boxes representing four orbitals of the outermost shell. The electrons in those orbitals are represented by dots, as before. A number of cxamples follow:

SIMPLIFIED NOTATION

Element	Electrons in outer shell	Symbolic representation of atom
Lithium	1	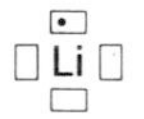 Li
Fluorine	7	 F
Oxygen	6	 O
Phosphorous	5	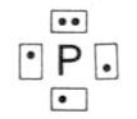 P
Neon	8	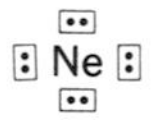 Ne
Calcium	2	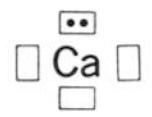 Ca
Chlorine	7	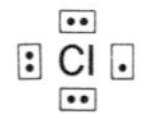 Cl

In this representation, one of the boxes represents an *s* orbital, and the other three represent *p* orbitals. The first two electrons are placed into the box representing the *s* orbital. Additional electrons are placed into the remaining boxes, until the octet is complete.

The first shell contains only one 1*s* orbital. The elements hydrogen and helium, for which the first shell is the outer shell, are therefore represented as follows:

H [·]   He [:]

The symbolism is readily extended to atoms that have gained or lost electrons to become ions. Thus:

$$[\ ]\,\overset{[\cdot]}{\underset{[\ ]}{\text{Na}}}\,[\ ] \xrightarrow{-e^{\ominus}} [\ ]\,\overset{[\ ]\oplus}{\underset{[\ ]}{\text{Na}}}\,[\ ] \quad \text{or} \quad \text{Na}^{\oplus}$$

The symbol at the right is a simpler version for the sodium ion. Other examples are:

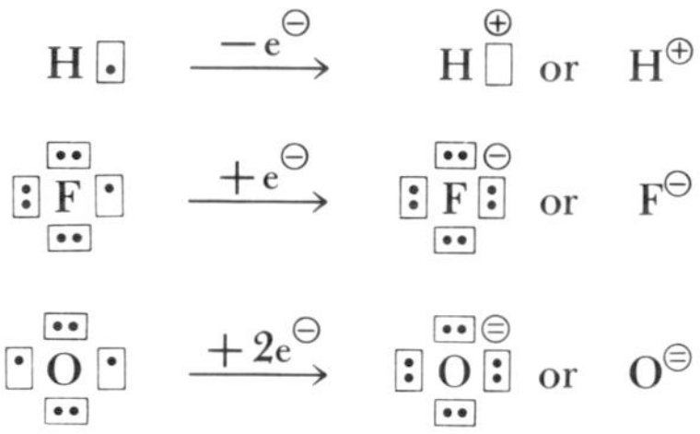

$$\text{H}\,[\cdot] \xrightarrow{-e^{\ominus}} \overset{\oplus}{\text{H}\,[\ ]} \quad \text{or} \quad \text{H}^{\oplus}$$

$$[:]\,\overset{[\cdot\cdot]}{\underset{[\cdot\cdot]}{\text{F}}}\,[\cdot] \xrightarrow{+e^{\ominus}} [:]\,\overset{[\cdot\cdot]\ominus}{\underset{[\cdot\cdot]}{\text{F}}}\,[:] \quad \text{or} \quad \text{F}^{\ominus}$$

$$[\cdot]\,\overset{[\cdot\cdot]}{\underset{[\cdot\cdot]}{\text{O}}}\,[\cdot] \xrightarrow{+2e^{\ominus}} [:]\,\overset{[\cdot\cdot]\ominus\!\!\!\!\ominus}{\underset{[\cdot\cdot]}{\text{O}}}\,[:] \quad \text{or} \quad \text{O}^{\ominus\!\!\!\!\ominus}$$

The plus and minus signs in these formulas indicate the number of electrons lost or gained.

Ionic compounds

The crystalline compound formed by the interaction of lithium metal with gaseous fluorine is called an *ionic* compound because it is composed of lithium ions and fluoride ions, held together by the mutual attraction of positive and negative charges. If our theory is correct, we can predict the formation of many other ionic compounds by the process of electron transfer. For example, calcium has two electrons in its outermost shell, and we can predict that it will lose them to become a calcium ion (Ca^{++}) if some *electron acceptor* is available. The electronic configuration of the chlorine atom suggests that it should react with calcium metal, and indeed it does. Furthermore, since each calcium atom must lose *two* electrons, whereas each chlorine atom accepts only one, the theory predicts that the compound formed from the interaction of the two elements contains *two* chlorine atoms for each calcium atom:

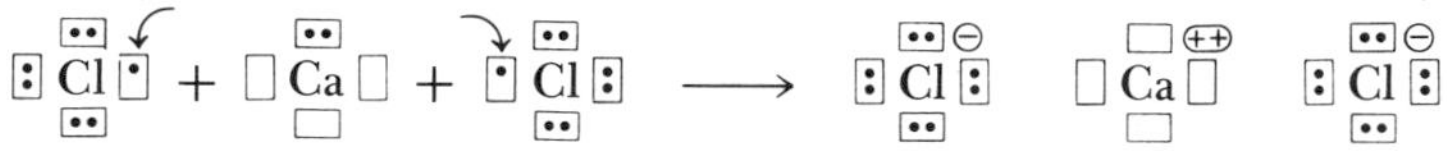

$$\mathrm{Cl} + \mathrm{Ca} + \mathrm{Cl} \longrightarrow \mathrm{Cl}^{\ominus} \quad \mathrm{Ca}^{\oplus\oplus} \quad \mathrm{Cl}^{\ominus}$$

Chemical analysis of the reaction product shows the ratio of calcium to chlorine to be $CaCl_2$, in accord with the theory.

When oxygen is brought into contact with calcium, the formation of an ionic compound is again anticipated, with each oxygen atom gaining the two electrons lost by each calcium atom.

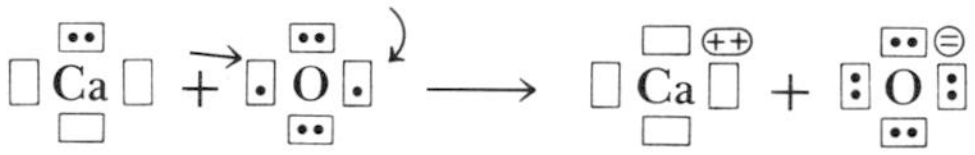

$$\mathrm{Ca} + \mathrm{O} \longrightarrow \mathrm{Ca}^{\oplus\oplus} + \mathrm{O}^{\ominus\ominus}$$

Since the number of electrons lost by the calcium metal (per atom) is equal to the number gained by the oxygen, the predicted formula is CaO, which again is in agreement with experiment.

Many examples of this kind can be furnished, in which the theory of electron transfer leads to the correct prediction of the atomic ratios in the compound. We therefore infer that electron transfer can indeed take place.

The outstanding and obvious characteristic of ionic compounds is that they are composed of charged particles (positive and negative ions). It is therefore possible to confirm the presence of ions through their electrical properties.

Electrical properties of ionic compounds

When current electricity was discovered around 1800, many people turned their attention to the effects that electric currents have on various substances. Most active in this field were Sir Humphry Davy, J. J. Berzelius, and Michael Faraday. Out of the experiments of these men developed the theory that the combination of elements to form compounds must somehow involve electrical charges. This idea is strikingly similar to our modern concept of ionic compounds, and it is interesting to note that it was formulated as early as 1815, long before our present-day ideas of atoms as consisting of electrically charged electrons and nuclei were even dreamed of. As a matter of fact, this theory was devel-

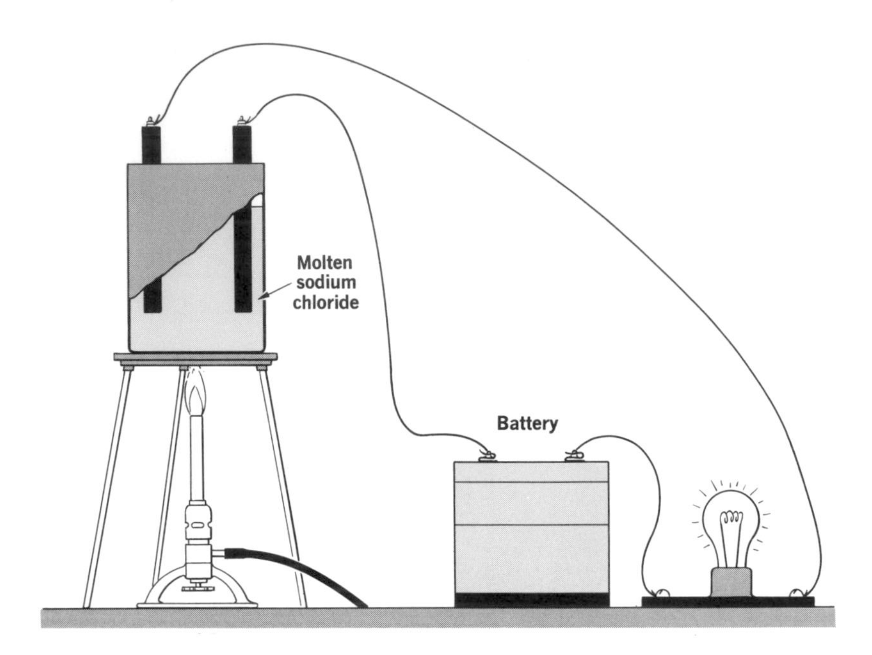

Fig. 12-1
Apparatus for electrolysis.

oped only a few years after the publication of Dalton's theory, which postulated the *existence* of atoms.

ELECTROLYSIS

One of the most striking properties of ionic compounds is their behavior when electric currents are passed through them. If solid sodium chloride is placed in a suitable container and heated to around 800°C, the salt melts and becomes liquid. If, now, a pair of graphite electrodes is placed in the melt and connected to a battery, as shown in Fig. 12-1, a series of events occurs that is invaluable in providing insight into the nature of ionic compounds. First of all, the molten salt conducts the current. This can be shown by inserting an electric lamp in the circuit: the lamp will glow and thus indicate that a current is flowing. On the other hand, if the salt is allowed to solidify again, the lamp will cease to glow and thus indicate that the current is no longer flowing. The experiment shows that the simple process of converting salt from a solid to a liquid will convert it from a nonconductor of electricity to a conductor.

Furthermore, while the current is flowing through the molten salt, things are happening at the electrodes which suggest that the flow of current must involve the ions of which the salt is assumed to be composed. At the negative electrode, the silvery metal, sodium, begins to deposit; while at the positive electrode, chlorine gas bubbles appear. In other words, the passage of the current is converting sodium chloride to the elements from which it was formed. This behavior is best explained by the assumption that molten sodium chloride consists of positively charged sodium ions and negatively charged chloride ions. Inasmuch as these ions are free to move, they are attracted to the electrode of opposite charge. That these ions exist *prior* to the melting process can be inferred from the fact that many other substances, on melting, do *not* conduct an electric current. Therefore the melting process itself does not create the ions; they must be present also in the solid sodium chloride. The solid does not conduct a current because the ions are held in a rigid crystal lattice and are not free to move to the electrodes.

In the molten salt, the positive sodium ions are attracted toward the negative electrode, and the negative chloride ions are attracted toward the positive electrode, as shown in Fig. 12-2. Each chloride ion that reaches the positive electrode deposits an electron there and becomes a

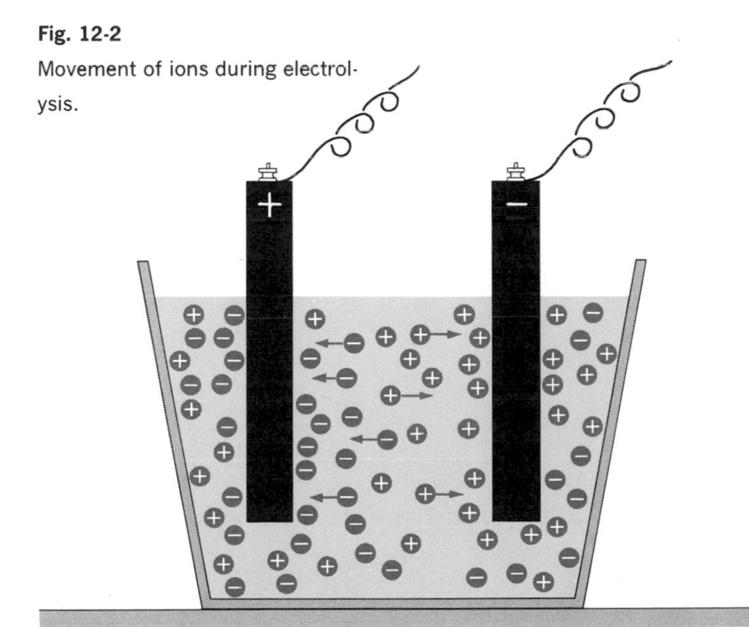

Fig. 12-2
Movement of ions during electrolysis.

neutral chlorine atom. Each sodium ion that reaches the negative electrode picks up an electron and becomes a neutral sodium atom. Thus, the net result of these changes is, obviously, the transport of electrons from the negative electrode to the positive electrode. This ionic conductance must be contrasted with the passage of electricity through a wire, in which case the electrons themselves move.

The deposition and acquisition of electrons is responsible for the chemical changes that occur at the electrodes. Thus, as the chloride ion loses an electron to the positive electrode, it becomes a neutral chlorine atom, and on its pairing up with another similarly formed chlorine atom there is formed a molecule of chlorine:

$$2\ \mathrm{Cl}^{-} \longrightarrow 2\ \mathrm{Cl} + 2e^{-}$$

Chloride ion — Chlorine atom

$$2\ \mathrm{Cl} \longrightarrow Cl_2$$

Chlorine gas

In a similar fashion, metallic sodium is formed from sodium ions:

$$\mathrm{Na}^{+} + e^{-} \longrightarrow \mathrm{Na}$$

Sodium ion — Metallic sodium

The process of decomposing compounds by means of an electric current is called *electrolysis*. It has wide industrial applications: in the commercial production of sodium, chlorine, hydrogen, copper, aluminum, and many other substances; in all types of electroplating, such as silver-plating or chromium-plating; and in chemical analysis.

Oxidation and reduction

We have seen several examples of processes in which atoms gain or lose electrons. Thus, when lithium and fluorine are brought into contact, the lithium atoms lose an electron and the fluorine atoms gain one. Similarly, during the electrolysis of molten sodium chloride, sodium ions gain an electron and chloride ions lose one. Processes in which electrons are gained are called *reduction,* and those in which electrons are lost are called *oxidation*. In the reaction of lithium with fluorine, the lithium is oxidized and the fluorine is reduced. In the electrolysis of sodium chloride, the chloride is oxidized to chlorine gas, and the sodium ions are reduced to metallic sodium.

These terms originated in the practice of metallurgy. When a pure metal reacts with oxygen, that is, when it is oxidized, it gives up electrons to the oxygen. When the oxide is reduced to the pure metal, the metal ions gain electrons.

Oxidation and reduction are of particular interest in biological chemistry. It is fair to say that life exists because the energy obtained from the sun is harnessed through an intricate series of oxidation-reduction reactions in the living organism.

The "molecular weight" of ionic compounds

In the case of ionic compounds, the concept *molecular weight* is not a simple one. To be specific, consider the results of studies of the vapor of the ionic compound, sodium chloride, at temperatures around 1000°C. At these temperatures sodium chloride is vaporized to a measurable extent. When the molecular weight of the vapor is determined, one discovers that the vapor consists not only of molecules with the formula Na^+Cl^-, but also of molecules having the formulas

$$\begin{matrix} Na^{\oplus}Cl^{\ominus} \\ Cl^{\ominus}Na^{\oplus} \\ (Na_2Cl_2) \end{matrix} \quad \text{and} \quad \begin{matrix} Na^{\oplus}Cl^{\ominus}Na^{\oplus} \\ Cl^{\ominus}Na^{\oplus}Cl^{\ominus}, \\ (Na_3Cl_3) \end{matrix} \quad \ldots \text{and so on}$$

In other words, while the *simplest* formula is NaCl, molecules with formulas that are integral multiples of this formula also exist, and it is meaningless to say which of these is the "true" formula of sodium chloride.

When we examine a solid crystal of sodium chloride—or of any other ionic compound, for that matter—the situation becomes even more complex, and we cannot possibly think in terms of molecules of sodium chloride. A tiny portion of a crystal of sodium chloride is shown diagrammatically in Fig. 12-3. True enough, in this crystal there is one sodium ion for each chloride ion, but we cannot say definitely which chloride ion belongs to which sodium ion. The crystal is really a network or lattice of sodium ions and chloride ions, the whole being held together by electrostatic attractions. In this sense, then, the whole crystal is really a single giant molecule.

This situation can be understood if we look more carefully at the properties of electrically charged bodies such as ions. If we take a positively charged body and bring a negatively charged body into its vicinity, the two bodies are attracted to one another, but it *does not* matter from which direction the first body is approached. The electrostatic force extends equally in all directions, and its strength depends only on the distance separating the two bodies. Thus, the number of negatively charged ions that can be attracted to a single positively charged ion depends, in the first approximation, only on the space available around the positive ion. In the case of a sodium chloride crystal (Fig. 12-3), each sodium ion is surrounded by six chloride ions and each chloride ion in turn is surrounded by six sodium ions.

There are other physical properties which indicate that compounds such as sodium chloride, calcium oxide, or silver bromide are ionic. For example, crystals of these substances are very difficult to vaporize. This indicates that the particles of which these crystals are composed are held together by very strong forces which are overcome only if considerable energy is supplied.

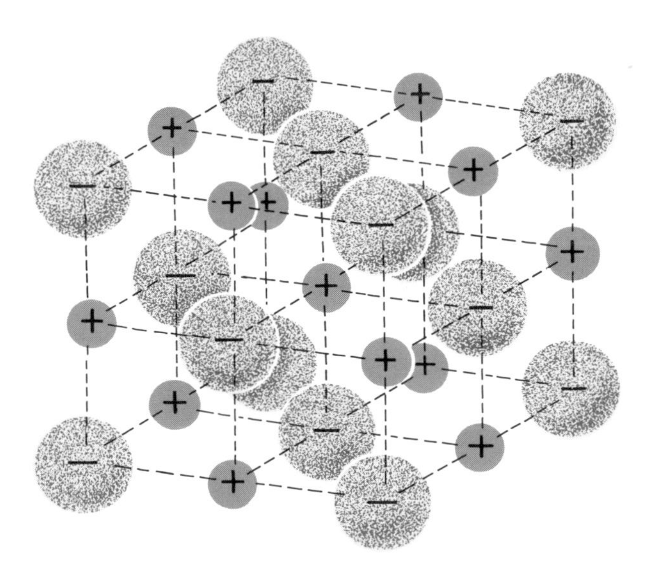

Fig. 12-3

Arrangement of ions in a sodium chloride crystal [Courtesy J. Quagliano, *Chemistry* (Englewood Cliffs, N. J.: Prentice-Hall, Inc., 1958), p. 42].

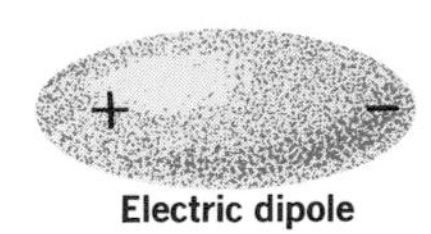

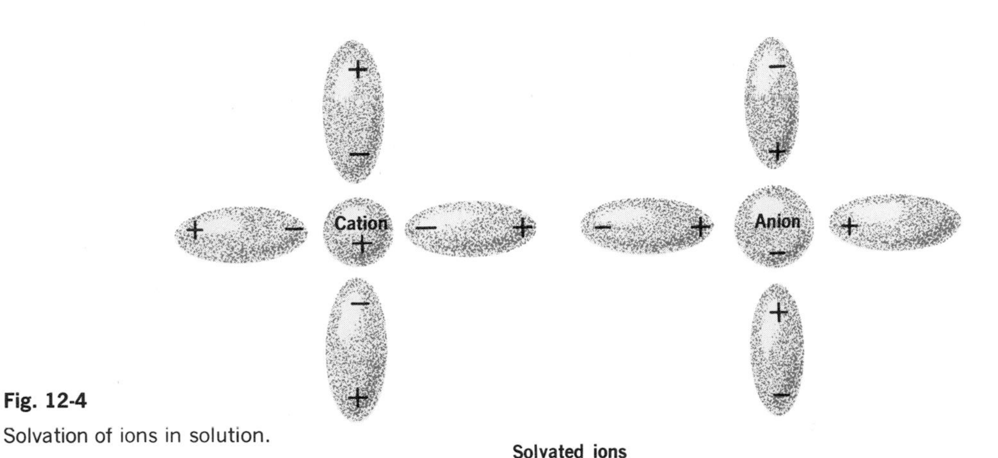

Fig. 12-4
Solvation of ions in solution.

Properties of ionic compounds in aqueous solution

In order for an ionic crystal to dissolve in water, strong forces are needed to overcome the mutual attraction of the ions in the crystal. Nevertheless, most salts in aqueous solution exist in the form of separated ions. Thus, solutions in water of lithium chloride (Li^+Cl^-), sodium chloride (Na^+Cl^-), and potassium chloride (K^+Cl^-) display identical chemical reactions, typical of the chloride ion, regardless of the nature of the positive ion. For example, they react with silver fluoride (Ag^+F^-) to produce the insoluble substance, silver chloride (Ag^+Cl^-). In each case, the reaction which takes place is that of silver ion with chloride ion:

$$Ag^{\oplus} \text{ (from } Ag^{\oplus}F^{\ominus}) + Cl^{\ominus} \text{ (from } Li^{\oplus}Cl^{\ominus}, Na^{\oplus}Cl^{\ominus}, \text{ or } K^{\oplus}Cl^{\ominus}) \longrightarrow AgCl \text{ (insoluble in water)}$$

Similarly, if a set of compounds contains a common positive ion, solutions of these compounds will display reactions characteristic of the common positive ion, regardless of the nature of the negative ions. For example, solutions of calcium chloride ($CaCl_2$), calcium bromide ($CaBr_2$), and calcium sulfide (CaS) all react with sodium carbonate (Na_2CO_3) to form insoluble calcium carbonate ($CaCO_3$).

There is much additional evidence for the view that ionic compounds dissociate into free ions in water solutions. For example, such solutions conduct electric current in much the same way as do the molten salts. The chemical reactions that occur at the electrodes again suggest that ions are being attracted to the electrodes.

The readiness with which ionic compounds dissolve in water can be well understood in terms of the structure of the water molecule. Anticipating a more detailed discussion (p. 231), we shall represent each water molecule as an electric *dipole*, that is, as having a positive charge at one end and an equal negative charge at the other. These dipoles interact strongly with the ions, eventually surrounding each ion with a sheath of dipoles, a process called *solvation* (see Fig. 12-4). The attraction of the dipoles for the ions lowers the energy of the ions and accounts for the readiness with which the ionic compound dissolves.

Suggestions for further reading

Fullman, Robert L., "The Growth of Crystals," *Scientific American*, March, 1955.

Germer, L. H., "The Structure of Crystal Surfaces," *Scientific American*, March 1965, p. 32.

House, J. E., "Ionic Bonding in Solids," *Chemistry*, *43*, Februrary 1970, p. 18.

Moore, W. J., *Seven Solid States* (New York: W. A. Benjamin Co., Inc., 1967).

Ryschkewitsch, George, *Chemical Bonding and the Geometry of Molecules* (New York: Van Nostrand Reinhold Company, 1963).

Sisler, Harry H., *Electronic Structure, Properties, and the Periodic Law* (New York: Van Nostrand Reinhold Company, 1963).

Updike, J., "The Dance of the Solids", (a poem) *Scientific American*, *220*, January 1969, p. 130.

Questions

1. How can one *predict* what the numerical value of the charge of a particular ion is most likely to be? What determines whether a particular element will form positive ions or negative ions? What is the nature of the force that holds two ions of opposite charge together?

2. How does the number of electrons lost (or gained) by an atom determine the formula of its ionic compounds?

3. Using the simplified notation introduced in this chapter, write electron-dot formulas for each of the following ions. Then predict whether the ion is likely to form stable ionic compounds.

(*a*) Al^{+++} (*b*) Mg^{+} (*c*) Na^{-}
(*d*) $S^{=}$ (*e*) F^{-} (*f*) Ca^{+++}

4. Write electron-dot representations of the following compounds. Show all charges on ions.
(*a*) SrO (*b*) NaI (*c*) Cs_2Sc

5. The formation of crystalline potassium chloride is accompanied by the evolution of a large amount of heat. Explain the nature of the process whereby potassium metal and chlorine gas are converted to the ionic solid, and explain the interactions that are responsible for the release of this large amount of energy.

6. Examine the periodic chart of the elements. Then list five elements that you would expect to donate electrons and five elements that you would expect to accept electrons when these elements react to form ionic compounds.

7. Balance the following equations using the method of algebra:
(*a*) $1\ CaCl_2 + a\ Na_2CO_3 \rightarrow b\ CaO + c\ CO_2 + d\ NaCl$
(*b*) $1\ Br_2 + a\ Li \rightarrow b\ LiBr$
(*c*) $1\ O_2 + a\ Mg \rightarrow b\ MgO$
(*d*) $1\ Fe_2O_3 + a\ MgCl_2 \rightarrow b\ FeCl_3 + c\ MgO$

8. Is there any limit on the number of negative ions that can bind with a positive ion?

9. Why can the concept of "molecular weight" not be applied in the case of ionic compounds? What concept can be substituted?

10. (*a*) Describe the process of *electrolysis* of molten salts.
(*b*) Write the equations for the reactions that take place at the electrodes, for the following molten salts: NaF; CaF_2; LiCl; KH.

11. Explain why an ionic compound such as sodium chloride dissolves with considerable ease in cold water, yet requires a temperature of over 800°C in order to melt.

12. What is the experimental evidence that free ions actually exist in solution?

13. After a salt is dissolved in water, what is the submicroscopic condition of the individual ions?

14.* From an examination of the diagram of ionization energies on page 173, suggest which of the noble gases would be most likely to form an ionic compound with fluorine.

15.* Describe two different kinds of crystal structures that are found for various alkali halides.

* Questions marked with an asterisk (*) are more difficult or require supplementary reading.

To him who is a discoverer in the field
(of science), the products of his
imagination appear so necessary and
natural that he regards them, and
would like to have them regarded by
others, not as creations of thought
but as given realities.

ALBERT EINSTEIN (1879–1955)

Thirteen

PRINCIPLES OF COMPOUND FORMATION: II. COVALENT COMPOUNDS

ALTHOUGH THE theory of the ionic bond will explain the stability of many known compounds, it cannot be the only explanation for molecule formation. This becomes clear as soon as we consider such molecules as H_2 or Cl_2, which consist of two *identical* atoms. Since in these molecules the two atoms have identical chemical properties, it is hard to see why one atom should gain an electron and the other lose one, as must be the case if the molecule is to consist of a pair of ions. Evidently some other mode of atomic bonding must exist.

Physical properties of covalent substances

The bonds in the molecules of hydrogen and chlorine are called *covalent* bonds, and substances with this kind of atomic bonding are called *covalent* substances. The nature of the covalent bond will be discussed in the next section. But first we shall describe certain macroscopic properties of covalent substances which indicate the absence of ions.

Melting point.—It was pointed out earlier that the ionic compound, sodium chloride, melts at 800°C. Solid chlorine, on the other hand, melts at −102°C (902° lower), and solid hydrogen melts at −259°C. The large decrease in melting point suggests that there is a fundamental difference in the melting process for ionic and covalent substances. In ionic crystals, melting actually involves the rupture of bonds between ions, since the orderly network of bonds between positive and negative ions that exists in the solid is disrupted on melting. It takes quite a bit of energy to rupture the ionic bonds, and the melting point of the crystal is therefore high. Conversely, if the melting point of a pure crystal is low, we may infer that ions are absent. In fact, the great majority of covalent substances have melting points below 300°C.

Behavior in the gas phase.—Covalent substances, in contrast to ionic ones, consist in the gas phase of definite molecules with a definite molecular formula. Thus, chlorine gas consists entirely of Cl_2 molecules, in contrast to gaseous sodium chloride, which consists of NaCl, Na_2Cl_2, Na_3Cl_3, etc.

Electrical conductivity.—The response of covalent substances to electrical currents is also quite different from that of ionic substances under similar conditions. Melts or solutions of ionic substances, it will be recalled, permit the passage of electric currents. On the other hand, covalent substances in the liquid state do not appreciably conduct electricity. Similarly, solutions of covalent substances in water are poor conductors (except when the covalent substance reacts with water to produce ions). The inability of covalent substances to conduct electrical currents is, further, a strong indication that charge-carrying ions are absent.

The covalent bond

Attempts at developing a theory of the covalent bond that did not involve the formation of ions began soon after J. J. Thomson's discovery of the electron. It was realized, particularly by G. N. Lewis, that atoms can achieve noble gas electron configurations and be held together by electrostatic forces without having to be converted to charged ions. The process by which this happens is best described as *electron sharing*.

A realistic description of electron sharing requires the concepts and language of wave mechanics. But before attempting such a description, let us examine the underlying electrostatic forces first.

Electrostatic forces.—As an example, consider the hydrogen molecule, H_2. This molecule consists of two nuclei (each bearing a charge

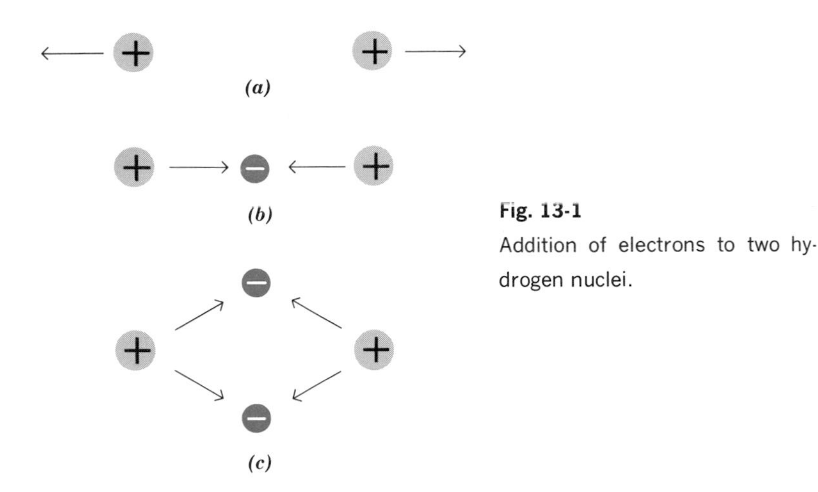

Fig. 13-1
Addition of electrons to two hydrogen nuclei.

of $+1$), and two electrons. The electrostatic forces acting in such a molecule will be:

— forces of attraction between each electron and each nucleus;
— forces of repulsion between the two nuclei, and likewise between the two electrons.

The effect of the electrons in holding together the nuclei is shown in Fig. 13-1. Imagine that we place the two nuclei (without the electrons) at a distance of separation equal to the normal internuclear distance in the hydrogen molecule, as in Fig. 13-1a. Because like charges repel, there will be a force of repulsion, and this system of particles will be inherently unstable. However, if we place an electron *between* the two nuclei, as in Fig. 13-1b, the two nuclei will be pulled toward the electron and the repulsion will be reduced or, perhaps, overcome. A second electron will enhance this effect (Fig. 13-1c). Thus, the two nuclei, by sharing the pair of electrons, will be bonded quite strongly. Other positions for the electrons are, of course, possible, because the electrons are in rapid motion; some of them are shown in Fig. 13-2. If, on the average, the electrons spend much of the time *between* the nuclei (or if there is at least one electron between the nuclei most of the time), a stable bond will be formed and the two hydrogen atoms exist as a stable H_2 molecule.

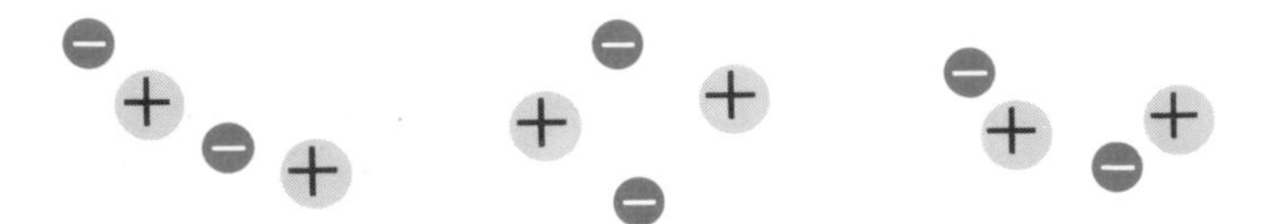

Fig. 13-2
Effect of electronic motion: Some typical arrangements of two electrons and two hydrogen nuclei.

Molecular orbitals.—These simple ideas involving electrostatic forces are consistent with the wave-mechanical model of the hydrogen molecule and can, in fact, be developed into a general wave-mechanical model for the formation of molecules, called *molecular orbital theory of covalent bonding*. The concept of an electron orbital is, of course, familiar from our previous study of (uncombined) atoms. Each electron orbital in an atom is characterized by a set of quantum numbers (n, l, m) and can hold two electrons of opposite spin. When the atom exists in its most stable or ground state, the electrons occupy the orbitals of lowest energy, so that the energy of the atom as a whole has the lowest possible value.

According to molecular orbital theory, when atoms combine to form a molecule, there is generated a new set of electron orbitals that is characteristic of the newly formed molecule. Each of these *molecular orbitals* can hold two electrons of opposite spin. And each molecular orbital (like an atomic orbital) depicts a region in space in which we are likely to find the electrons that occupy the orbital.

Wave mechanics enables us to derive the molecular orbitals from the electron orbitals of the atoms that combine. Although the theory is not yet complete, the basic ideas are well established. We wish to apply them to describe the formation of a hydrogen molecule from two hydrogen atoms.

Hydrogen molecule.—Recall that for the hydrogen atom in the ground state, the electron is in a 1s orbital. If two such orbitals are added together, there is produced a new orbital that is common to both atoms, and which is, therefore, a molecular orbital of the hydrogen molecule. This statement will become more concrete by reference to Fig. 13-3, which is based on wave-mechanical calculations. The figure shows cross-sections through the centers of two separated hydrogen atoms whose electrons are in 1*s* orbitals, and of the hydrogen molecule formed from them. The density of the stippling at any point is a measure of the probability of finding the electron in a small fixed volume around the point. The greater the density, the greater the probability. In the separated hydrogen *atoms,* the stippling is symmetrical about the nucleus, showing that there is an equal probability of finding the electron along any line through the nucleus. In the hydrogen *molecule,* however, that symmetry is lost, and the densely stippled area around each nucleus has shifted toward the other nucleus. This means that there is now a fairly high probability of finding an electron in the region between the two nuclei. Thus, if we could take a random series of instantaneous snapshots of the molecule, we would find that in most snapshots, at least one electron exists in the region between the two

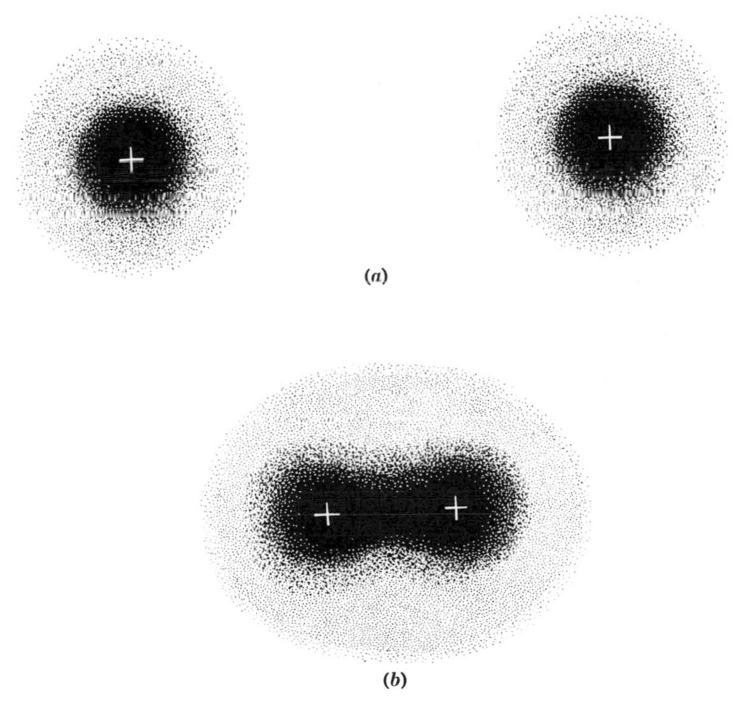

Fig. 13-3
Electron distribution in (a) separated hydrogen atoms and (b) a hydrogen molecule.

nuclei; and the average electrostatic force, based on all snapshots, would be decidedly one of attraction.

Although the molecular orbital of the hydrogen molecule does not have the spherical symmetry of the 1*s* orbitals of the separated atoms, it is nevertheless quite symmetrical, showing perfect symmetry with respect to the two hydrogen nuclei. This kind of symmetry implies that the electrons in the molecular orbital are shared equally by both atoms. It is no longer possible to associate either electron with either nucleus: both electrons must be assigned to both nuclei. As a result, each hydrogen atom acquires a share in a stable helium-like electron configuration.

The helium-like stability of the electron configuration becomes even more clear if we represent the formation of the hydrogen molecule in our previous notation. As the two hydrogen atoms approach each other,

H [•] →   ← [•] H

the two half-filled 1*s* orbitals merge or overlap,

H[•[]•]H

and a single molecular orbital, holding both electrons, results,

H [••] H   (Compare this with He [:])

Chlorine Molecule.—The formation of a chlorine molecule from two chlorine atoms is more complicated, but its essential aspects can be described in the same way. The electron configuration of the chlorine atom is shown below, both in detail and in simplified notation.

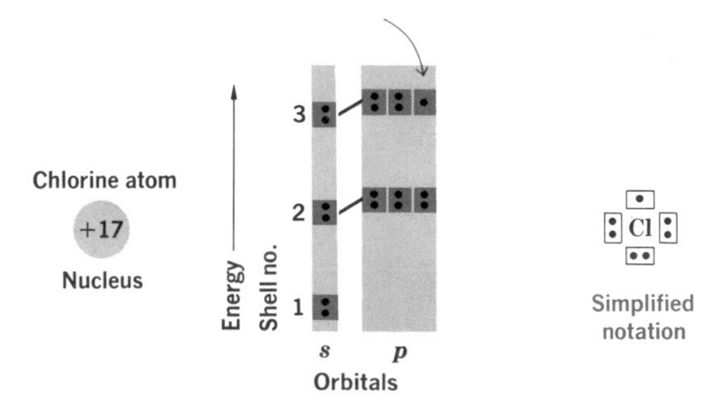

As shown by the arrow, one of the $3p$ orbitals is half-filled. When two chlorine atoms approach each other,

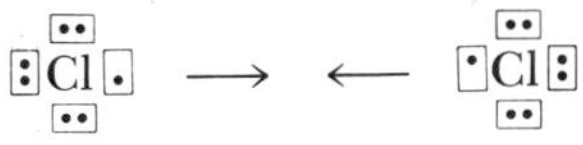

```
 ••            ••
:Cl·  ⟶  ⟵  ·Cl:
 ••            ••
```

the two half-filled $3p$ orbitals merge or overlap,

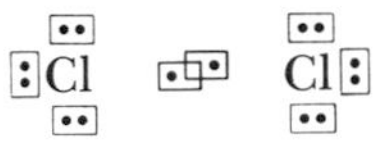

```
 ••         ••
:Cl   ·⊡·   Cl:
 ••         ••
```

and a single molecular orbital, holding both electrons, results.

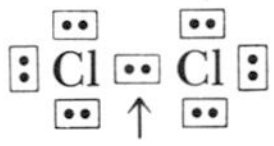

```
 ••     ••
:Cl  ••  Cl:
 ••  ↑  ••
```

The electrons in the new molecular orbital are shared equally by both chlorine atoms. As a result, each chlorine atom acquires a stable octet of electrons.

It is a triumph for this theory that it explains why the molecules of hydrogen and chlorine are diatomic: when each atom in the molecule has acquired a stable noble gas electron configuration, there remains little tendency for further bond formation. Thus the diatomic Cl_2 molecule shows little tendency to combine with another chlorine atom to form Cl_3, or with a Cl_2 molecule to form Cl_4.

Polar covalent bonds: the hydrogen chloride molecule.—It is not necessary, in order for covalent bonds to form, that the atoms sharing the electron pair be identical. As an example, let us consider the hydrogen chloride molecule. The substance, hydrogen chloride, has a low boiling point (−85°C), and the liquid is a poor conductor of electricity. These properties suggest that the atoms are covalently bonded, with each atom achieving a stable electron configuration through sharing.

In our simplified notation, the formation of a covalent hydrogen chloride molecule from the separated atoms is represented as follows:

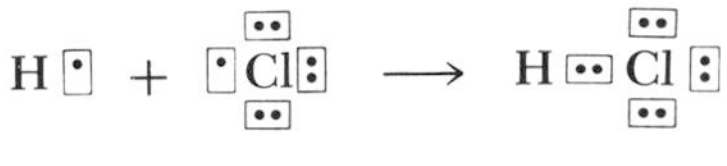

```
         ••               ••
H·  +  ·Cl:   ⟶   H ••  Cl:
         ••               ••
```

Note that the hydrogen atom acquires a stable helium-like electron configuration, and the chlorine atom acquires a stable octet. However, because the two atoms are dissimilar, we should not expect the pair of electrons to be shared equally between them.

This expectation is borne out by an investigation of the *electric dipole moment* of the molecule. The dipole moment is a measure of the asymmetry of the distribution of electrical charge. If one end of the molecule has, on the average, a positive charge, while the other end is negatively charged, the molecule is said to have a dipole moment. Measurement shows that the hydrogen molecule, where the atoms are identical, has no dipole moment, but that hydrogen chloride does. This strongly suggests that the electrons in the HCl molecule are not shared equally between the two atoms. Since we know that the chlorine atom has a strong tendency to form the negative chloride ion, it is probable that the shared electron pair is actually displaced towards the chlorine atom, making that end of the molecule somewhat negative.

H :Cl: or (+−)

Covalent bonds between dissimilar atoms commonly show this type of unequal charge distribution. Consequently, the molecules of many covalent compounds have substantial dipole moments. Such molecules are referred to as *polar molecules*. Some familiar examples are the molecules of water and ammonia.

Covalent molecules consisting of more than two atoms

We have seen that atoms can achieve stable electron configurations by electron sharing, and that the sharing of electrons by two atoms leads to a stable bond. These concepts will now be applied to predict covalent bonding in more complex molecules.

Localized bonding orbitals.—A complex molecule contains many electrons and, accordingly, has many filled electron orbitals. A *localized bonding orbital* is a molecular orbital in which a pair of electrons is shared by two atoms and forms a bond between them. The localized bonding orbitals of complex molecules are quite similar to the bonding orbitals of diatomic molecules: a localized bonding orbital results from the merging, or overlap, of two atomic orbitals belonging to two adjacent atoms. The electrons occupying the localized bonding orbital will have an increased probability of existing in the region between the two atomic nuclei, and only a small probability of existing elsewhere in the molecule, outside the two atoms that they hold together. Thus the electrons are *localized,* and the resulting covalent bond is called a *localized bond.*

How many bonds does an atom form?—If we assume that the atoms in covalent molecules are joined by localized bonds only, then an atom can form as many covalent bonds as it has half-filled outer orbitals. For example, we saw that hydrogen and chlorine atoms each have *one* half-filled outer orbital, hence will form only *one* covalent bond.

H [•]   [:] Cl [•] (with [••] above and below Cl)

Oxygen atoms, on the other hand, have *two* half-filled orbitals; hence, they can form *two* localized covalent bonds.

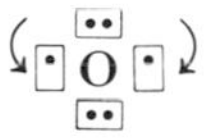

[•] O [•] (with [••] above and below O)

An oxygen atom can, therefore, combine with two hydrogen atoms to form a molecule with the formula H_2O, in which both hydrogen atoms are joined to a central oxygen atom. All three atoms thus acquire a stable noble gas electron configuration.

H [•] + [•] O [•] + [•] H ⟶ H [••] O [••] H (with [••] above and below each O)

Water

Physical measurements have shown that the charge distribution is such that the oxygen atom is somewhat negative; that is, the water molecule is polar.

Similarly, nitrogen atoms, with three half-filled orbitals, can form three localized bonds.

```
  [••]
[•] N [•]
  [•]
```

If the nitrogen atom combines with three hydrogen atoms, a molecule with the formula NH_3 is formed, in which all three of the hydrogen atoms are bonded to the nitrogen atom.

```
              [••]                                [••]
H [•]  +  [•] N [•]  +  [•] H   ——→   H [••] N [••] H
              [•]                                 [••]
               +                                   H
              [•]
               H                           Ammonia, NH3
```

And in fact, this is the correct formula for the familiar substance, ammonia.

The electron configuration of the carbon atom in its ground state is $1s^2\ 2s^2\ 2p^2$. As has been pointed out in Ch. 11, the energy levels of the orbitals can be represented as follows:

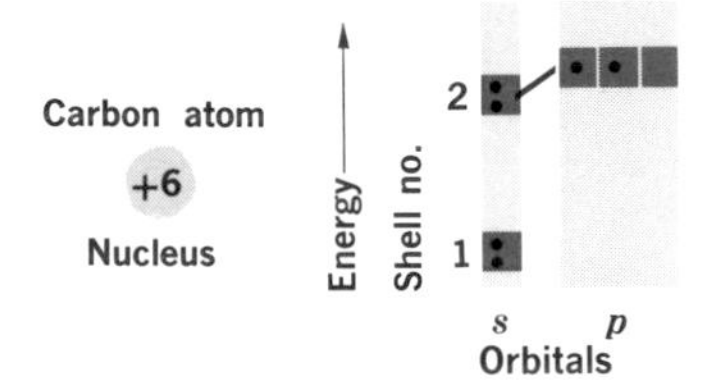

Since this diagram shows the carbon atom as having two half-filled orbitals, it suggests that carbon should form two covalent bonds under normal circumstances. However, investigation of a large number of carbon-containing compounds shows that the carbon atom usually forms four covalent bonds. Furthermore, these bonds are identical when the carbon atom is bonded to four identical atoms, as in CH_4 or CCl_4. To account for these facts, we first postulate that one of the $2s$ electrons is moved to a $2p$ orbital. This gives us four half-filled orbitals.

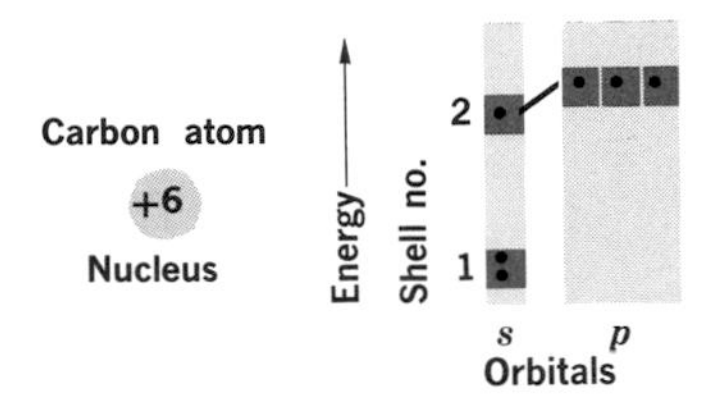

It is expected that such an atom will form four localized covalent bonds. However, those bonds will not all be identical, since one bond utilizes an *s* orbital whereas the others utilize *p* orbitals. To account for the observed equivalence of the bonds, we must assume that a further change takes place. This change is called *bond hybridization* and involves a rearrangement of the electron distribution so that four equivalent orbitals result. For purposes of bond formation, the carbon atom can therefore be represented as having four identical half-filled orbitals in its outer shell.

[·]
[·] C [·]
[·]

Hybridized carbon atom

When such an atom combines with, say, four hydrogen atoms, all four of the outer orbitals are involved in bond formation.

H
[:]
H [··] C [··] H
[:]
H

Methane, CH_4

The formula, CH_4, correctly represents the molecules of the known substance methane, which is a principal component of natural gas. Further discussion of bond hybridization will be postponed until the next chapter.

It is possible to write down a literally endless number of molecular formulas in which the atoms are joined by covalent bonds. Some additional formulas, showing molecules of somewhat greater complexity, are listed in the following table. The table also gives the names and boiling points of actual compounds having these formulas.

SOME COVALENT MOLECULAR FORMULAS

Formula	Actual compound
H [:] [··] H [··] C [··] Cl [:] [:] [··] H	CH_3Cl, Methyl chloride b.p. −24.2°C
[··] [:] Cl [:] [··] [:] [··] [:] Cl [··] C [··] Cl [:] [··] [:] [··] [:] Cl [:] [··]	CCl_4, Carbon tetrachloride b.p. 76°C
H H [:] [:] H [··] C [··] C [··] H [:] [:] H H	C_2H_6, Ethane b.p. −88.3°C
[··] [··] H [··] N [··] N [··] H [:] [:] H H	N_2H_4, Hydrazine b.p. 113.5°C
H H H H H [:] [:] [:] [:] [:] H [··] C [··] C [··] C [··] C [··] C [··] H [:] [:] [:] [:] [:] H H H H H	C_5H_{12}, *Normal* pentane b.p. 36.2°C

The theory that atoms can acquire stable electron configurations through sharing of electrons with other atoms is well established. In the great majority of cases, if one writes a formula that looks stable (each atom has a noble gas electron configuration), a corresponding actual substance almost invariably exists in nature or can be prepared in the laboratory. On the other hand, if the formula is such that some of the atoms do not have a noble gas configuration, a corresponding actual substance is either not known, or else is highly reactive.

For example, it is known that a substance with the formula CH_3 is highly reactive. This reactivity is suggested by the formula:

```
    H
   [:]
H [··] C [·]
   [:]
    H
```

Here, the carbon atom still has a half-filled orbital that can be used to form a covalent bond to some other atom or group of atoms. Thus, two CH_3 units can couple according to the equation:

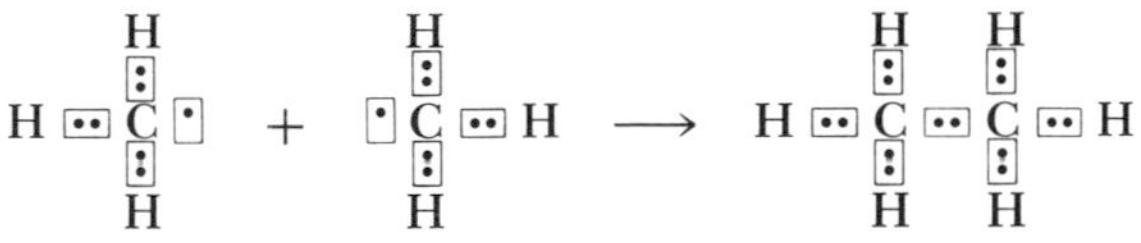

Ethane

Double bonds and triple bonds

It is possible, in principle, that two atoms be held together by more than a single covalent bond. Nitrogen molecules (N_2) are a case in point. The nitrogen atoms achieve a stable octet of electrons by the sharing of three electron pairs, to form a *triple* bond:

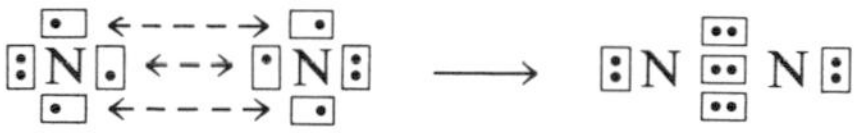

2 Nitrogen atoms

Nitrogen molecule; the atoms are joined by a triple bond

Similarly, in carbon dioxide (CO_2) each atom achieves a stable octet if the carbon atom is joined to the two oxygen atoms by double bonds:

Double bonds

The formation of double or triple bonds between two atoms is not at all uncommon. Since the double bond involves the sharing of two electron pairs, it is not surprising that the energy required to separate the two atoms is nearly twice as great as in the case of a single bond. Analogously, to separate a pair of atoms joined by a triple bond would require nearly three times as much energy. The increase in the number of shared electron pairs between two atoms is always accompanied by a decrease in the distance between the bonded atoms.

A much-studied example to illustrate these points is the series C_2H_6, C_2H_4, and C_2H_2. The formulas, chemical names, and other properties of the corresponding compounds are listed in the table.

SINGLE, DOUBLE, AND TRIPLE BONDS

Formula and name	Boiling point	Energy of the bond between the carbon atoms (calories)	Carbon-carbon bond distance (cm)
C_2H_6, ethane	−88.3°C	80,000	1.54×10^{-8}
C_2H_4, ethylene	−103.9°C	145,000	1.33×10^{-8}
C_2H_2, acetylene	−88.5°C	198,000	1.20×10^{-8}

Note that the bonding of the carbon atoms changes from a single bond in C_2H_6 to a triple bond in C_2H_2, and that the molecules depicted by these formulas are capable of stable existence. Each carbon atom is surrounded by an octet of electrons. The strength of the bond joining the two carbon atoms is measured by the *bond energy*, which is the amount of energy required to break the bond. The data in the table show the progressive increase in bond energy from single bond to double bond to triple bond. They also show the decrease in the distance between the centers of the carbon atoms.

Elements that form covalent bonds

We have seen that it is possible to determine experimentally whether a given compound is ionic or covalent, because of the sharp contrast in the properties of the two types of compounds. The question now arises whether it is possible to *predict* the nature of the bonding in a given compound.

In seeking an answer to this question, we recall that the covalent bond was first conceived to explain the bonding of *identical* atoms (as in

H_2 or Cl_2), and that the ionic bond was postulated for compounds of such dissimilar elements as the alkali metals and the halogens. In terms of the periodic chart (page 171), ionic compounds are likely to be formed between the metallic elements on the left of the chart and the nonmetallic ones on the right—excepting, of course, the noble gases. For the main-group elements, covalent bonds are formed:

1. between identical atoms.
2. between atoms belonging to the same column of the periodic chart, such as iodine and bromine.
3. between atoms belonging to adjacent columns, such as aluminum and carbon.
4. between any two nonmetallic atoms, such as carbon and oxygen, hydrogen and nitrogen, or boron and chlorine.

Other types of bonds

THE COORDINATE-COVALENT BOND

The coordinate-covalent bond is an interesting variation on the localized covalent bond in which one atom supplies *both* of the bonding electrons. A case in point is the compound formed between ammonia (NH_3) and boron trifluoride (BF_3). The structure of the ammonia molecule has already been discussed; it is as follows:

```
    [••] <——— Unshared electron pair
H [••] N [••] H
    [:]
     H
```

Note that three electron pairs are used to form the N–H bonds, but that the fourth electron pair is not being shared by any other atom.

Now let us consider the bonding in boron trifluoride:

```
  [••]           [  ]           [••]                          Vacant orbital
[:] F [•]  +  [•] B [•]  +  [•] F [:]                   [••]  [  ]  [••]
  [••]           [•]            [••]         ——>     [:] F [••] B [••] F [:]
                  +                                     [••]  [:]  [••]
                 [•]                                        [:] F [:]
              [:] F [:]                                       [••]
                 [••]
```

Originally, the boron atom has three outer electrons. Through sharing of electrons with the fluorine atoms it can acquire a share in six electrons, but it still needs two electrons to complete the octet. These two electrons can be furnished by the ammonia molecule:

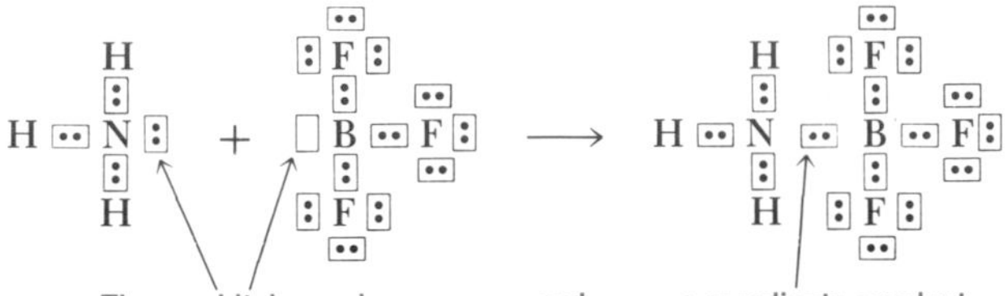

$H_3N: + BF_3 \longrightarrow H_3N:BF_3$

These orbitals overlap and a coordinate-covalent bond results.

The coordinate-covalent bond formed by the blending of the two atomic orbitals is just like an ordinary covalent bond, except that the nitrogen atom has supplied *both* electrons.

In this connection, it is interesting to note that boron trifluoride is a moderately stable compound, even though the boron atom does not have an octet of electrons. Substances whose formation violates the octet rule are occasionally found. But such substances generally react further to form substances in which the octet rule is satisfied.

COMPOUNDS WITH BOTH IONIC AND COVALENT BONDS

So far in our discussion of bonding, we have confined ourselves to molecules in which there are *only* ionic bonds or *only* covalent bonds. There are, of course, numerous substances in which bonds of both types exist simultaneously.

Potassium hydroxide is an example. The molecule, KOH, contains one ionic bond and one covalent bond. First the hydrogen atom is bonded to the oxygen atom by a covalent bond to form the unit:

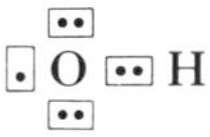

$\cdot\ddot{\underset{\cdot\cdot}{O}}:H$

Then the potassium atom transfers an electron to form the stable molecule:

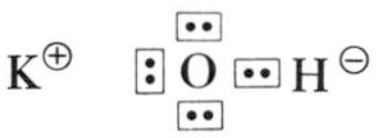

$K^{\oplus}$ $[:\ddot{O}:H]^{\ominus}$

Thus, when potassium hydroxide is dissolved in water, it dissociates into potassium ions (K^+) and hydroxide ions:

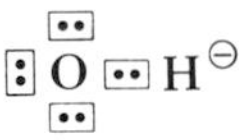

$[:\ddot{O}:H]^{\ominus}$ or simply OH^-

A second example is furnished by silver nitrate, whose formula is $AgNO_3$. The three oxygen atoms are bonded to the nitrogen to form the unit NO_3, to which a silver atom transfers an electron to form NO_3^-:

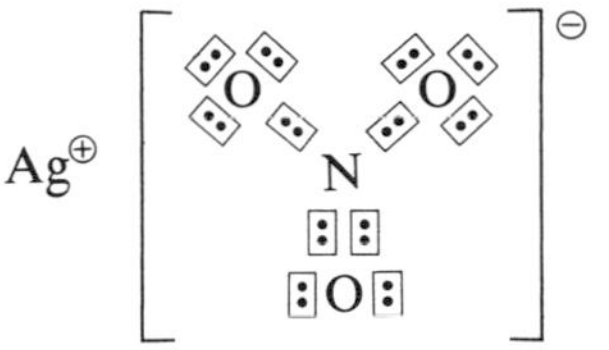

$Ag^{\oplus}$ $[NO_3]^{\ominus}$ (O, O, N, O electron-dot structure)

Solutions of silver nitrate are thus composed of silver ions (Ag^+) and nitrate ions (NO_3^-). Many other common substances exhibit this kind of dual bonding, some of which have been previously mentioned. For example, calcium sulfate ($Ca^{++}SO_4^{=}$), potassium carbonate [$(K^+)_2CO_3^{=}$], and sodium azide ($Na^+N_3^-$).

Suggestions for further reading

Barrow, G. M., "The Structure of Molecules" (New York: W. A. Benjamin, Co., Inc., 1963).

Ferreira, R., "Molecular Orbital Theory, an Introduction," *Chemistry, 41,* June 1968, p. 8.

Gray, H. B., *Electrons and Chemical Bonding,* (New York: W. A. Benjamin Co., Inc., 1964).

Herz, W., *The Shape of Carbon Compounds* (New York: W. A. Benjamin Co., Inc., 1963).

Hochstrasser, R. M., *The Behavior of Electrons in Atoms,* (New York: W. A. Benjamin Co., Inc., 1964).

Questions

1. Compare the behavior of covalent compounds with that of ionic compounds in the gas phase. Why does the behavior of covalent compounds suggest the absence of ionic charges?

2. What is a molecular orbital? How do molecular orbitals differ from atomic orbitals?

3. What is the nature of the forces that are responsible for covalent bonding?

4.* If the hydrogen molecular orbital contains only one electron, the species that results may be represented by the symbol $(H \cdot H)^+$, or H_2^+. Would you expect the one-electron bond in H_2^+ to be a stable bond? In the light of your answer, discuss the statement: "A covalent bond requires the sharing of a pair of electrons."

5. How do you predict the number of covalent bonds that a given atom is likely to form?

6. Give an example of:
(*a*) An atom that is likely to form one covalent bond only.
(*b*) A family of elements whose atoms are likely to form three covalent bonds.

7. Explain the origin of the dipole moment in such covalent molecules as HCl and HOH.

8. What is meant by the term "bond hybridization?"

9. Write plausible electron-dot formulas (in which each atom acquires a noble gas electron configuration) for the following molecules:

(*a*) NI_3 (*b*) C_3H_8
(*c*) C_3H_6 (*d*) H_2SO_4
(*e*) H_3PO_4 (*f*) C_4H_{10}
(*g*) C_4H_8 (*h*) C_4H_6
(*i*) H_2O_2 (*j*) HOCl
(*k*) NH_2^- in $K^+NH_2^-$ (*l*) N_2O_4

10. Why doesn't a stable covalent molecule exist with the structure

$$\mathrm{Na} : \overset{\cdot\cdot}{\underset{\cdot\cdot}{\mathrm{Cl}}} : \quad ?$$

11. Try writing electron-dot formulas for the following molecules: (*a*) NO; (*b*)

* Questions marked with an asterisk (*) are difficult or require supplementary reading.

XeF_4; (*c*) N_2O; (*d*) CH_2. Which molecules are likely to correspond to stable substances? Which molecules contain coordinate covalent bonds?

12. Represent, by electron-dot formulas, the following reactions:
(*a*) $H_2 + C_2H_4 \longrightarrow C_2H_6$
(*b*) $N_2H_4 + 2BH_3 \longrightarrow B_2N_2H^{10}$ (with two B–N bonds).

One of the most striking evidences of the reliability of the chemist's methods of determining molecular structure is the fact that he has never been able to derive satisfactory structures for supposed molecules which are in fact nonexistent.

LOUIS P. HAMMETT (1894–)

Fourteen

THE ARCHITECTURE OF COVALENT MOLECULES

OUR KNOWLEDGE concerning the nature of the chemical bond enables us to describe the structure of molecules in considerable detail. By structure we mean the organization of the atoms within a given molecule. Previously we had concerned ourselves chiefly with the *composition* of molecules. But knowing the composition of a molecule is not the same as knowing its structure, any more than knowing the quantities of materials involved in a building is the same as knowing its architecture.

A most satisfying chapter in the history of chemistry has been the growth in our knowledge of the structure of even the most complex molecules. Substances whose names are ordinary household words, such as penicillin, chlorophyll, or D.N.A., consist of molecules composed of dozens of atoms, and our knowledge of the architecture of these molecules is quite exact. Even such extremely complex substances as the proteins are yielding the secrets of their construction.

The diagrams of molecular structure that will be presented in this and the following chapter are remarkably detailed when one considers that all the information upon which they are based is quite indirect.

The inadequacy of the molecular formula

In the preceding chapters, we considered the forces binding atoms together to form molecules. We learned that there are two main types of chemical bonding, the ionic bond and the covalent bond, and that these lead to quite different results. Ionic bonding results in the formation of large crystalline aggregates of indefinite size. On the other hand, covalent bonding leads to the formation of definite molecular units in which specific numbers of atoms are joined to form discrete molecules. These molecular units can be represented by molecular formulas, such as CH_4, H_2O, or Cl_2, in which the elements are written in a completely arbitrary order, regardless of how they are actually joined in the molecule. The information conveyed by this type of representation is merely the number of atoms that have combined to form one molecule of the substance. The molecular formula tells us nothing about the way in which the atoms are linked.

But the details of the construction of the molecule are important—a fact that was appreciated early in the development of chemistry. The molecular formula, which tells us nothing about the molecular architecture, is therefore inadequate. One of the simplest ways of appreciating this fact is to consider that two entirely different substances can have exactly the same number and kind of atoms in their molecules. For example, the two substances, dimethyl ether and ethyl alcohol, both have the molecular formula C_2H_6O. Ethyl alcohol is the active ingredient in alcoholic beverages; dimethyl ether is chemically similar to the ether used in surgical anesthesia. Ethyl alcohol is a liquid boiling at 78 °C and is completely soluble in water; dimethyl ether is a gas at room temperature and is only slightly soluble in water. Ethyl alcohol reacts with metallic sodium; dimethyl ether does not. Clearly, these two materials are different substances, even though their molecular formulas are identical.

Structural isomerism

Substances such as ethyl alcohol and dimethyl ether, with the same molecular formula but different physical and chemical properties, are called *isomers*.† Isomerism can arise whenever it is possible to build up molecules of different architecture from the same set of atoms without violating the octet rule.

It will be recalled that covalently bonded molecules are capable of stable existence if each atom has acquired a noble gas electron configuration. *This means that every possible arrangement of the atoms in which the octet rule is satisfied represents potentially a stable substance.* If it is possible to arrange the atoms in only one way while satisfying this requirement, then only one substance with this molecular formula is expected to exist. But if there are several alternative arrangements, several substances with this molecular formula are anticipated.

Let us now consider in detail the two isomeric substances with the molecular formula C_2H_6O: ethyl alcohol and dimethyl ether. Molecules with the formula C_2H_6O consist of nine atoms of three different elements. The possible number of different arrangements of these atoms is very large if no attention is paid to the octet rule. But only two arrangements conform to the rule. They are as follows:

```
      H  H                       H       H
      |  |                       |       |
  H—C—C—Ō—H               H—C—Ō—C—H
      |  |                       |       |
      H  H                       H       H

        I.                           II.
```

In writing structural formulas I and II, we have adopted a convenient shorthand. Filled two-electron orbitals, which up to now have been represented by a box containing two dots, are simply indicated by a dash.

It is gratifying to know that the number of actual substances with the formula C_2H_6O is exactly equal to the number of stable structural formulas that one can write for molecules with this composition. There are two substances and two formulas. Assuming that these formulas actually represent the molecular architecture of ethyl alcohol and dimethyl ether, the problem then remains of assigning the appropriate structural formula to the right substance.

† Derived from the Greek roots *iso* (equal) and *meros* (part), i.e., the molecules of the two substances are constructed from the same (atomic) building materials, and in the same proportions.

ASSIGNMENT OF STRUCTURAL FORMULA

This problem can be solved because *molecular structure has chemical consequences*. A chemical reaction involves the making and breaking of bonds between atoms. Since the structural formula depicts the bonds in the molecule, it provides information about the reactions that a substance with such a structure would be expected to undergo.

How do the bonds in structure I differ from those in structure II? Structure I contains an O—H bond, whereas structure II does not. In structure I, the two carbon atoms are linked directly, whereas in structure II they are separated by an oxygen atom. Both structures contain several C—H bonds and at least one C—O bond.

When we know the chemical reactions of each of the two substances, we will be able to assign the appropriate structural formula to each substance by taking advantage of these differences. For example, one and only one of the two substances is expected to exhibit the reactions characteristic of an O—H bond. Let us find out whether this is in fact the case. The first question that we must answer is: What *are* the chemical properties of an O—H bond? Common sense would dictate that we can answer this question by studying the chemical properties of a molecule in which *all* the bonds are O—H bonds. The water molecule is such a molecule. One of the characteristic properties of water molecules is that they react readily with alkali metal atoms; hydrogen gas is evolved, and an alkali hydroxide, such as NaOH, is formed.

$$2\,Na \quad + \quad 2\,H—\underline{\overline{O}}—H \quad \longrightarrow \quad 2\,NaOH \quad + \quad H_2$$

The ready reaction with alkali metals is characteristic of the O—H bond but not of the C—H bond. Methane (CH_4), which has only C—H bonds, does not react readily with sodium.

Since one of the two isomeric substances, ethyl alcohol, reacts with sodium in the manner characteristic of an O—H bond, while the other, dimethyl ether, does not, ethyl alcohol must be assigned structure I. This leaves structure II to be assigned to the other isomeric substance, dimethyl ether.

The accuracy of these assignments can be tested further by carrying out a chemical reaction, the products of which can be predicted correctly only if the structures are assigned correctly. A suitable reaction is that of ethers and alcohols with hydrogen iodide. For structures I and II, the predicted reactions are as follows:

STRUCTURE I:

$$\begin{array}{ccccccc} & \mathrm{H} & \mathrm{H} & & & & \\ & | & | & & & & \\ \mathrm{H}- & \mathrm{C}- & \mathrm{C}- & \underline{\overline{\mathrm{O}}}-\mathrm{H} & + & \mathrm{H}-\underline{\overline{\mathrm{I}}}| & \longrightarrow \\ & | & | & & & & \\ & \mathrm{H} & \mathrm{H} & & & & \end{array} \begin{array}{cccccc} & \mathrm{H} & \mathrm{H} & & & \\ & | & | & & & \\ \mathrm{H}- & \mathrm{C}- & \mathrm{C}- & \underline{\overline{\mathrm{I}}}| & + & \mathrm{H}-\underline{\overline{\mathrm{O}}}-\mathrm{H} \\ & | & | & & & \\ & \mathrm{H} & \mathrm{H} & & & \end{array}$$

Hydrogen iodide

Ethyl iodide

STRUCTURE II:

$$\begin{array}{ccccccc} & \mathrm{H} & & \mathrm{H} & & & \\ & | & & | & & & \\ \mathrm{H}- & \mathrm{C}- & \underline{\overline{\mathrm{O}}}- & \mathrm{C}-\mathrm{H} & + & 2\mathrm{H}-\underline{\overline{\mathrm{I}}}| & \longrightarrow \\ & | & & | & & & \\ & \mathrm{H} & & \mathrm{H} & & & \end{array} \begin{array}{ccccc} & \mathrm{H} & & & \\ & | & & & \\ 2\mathrm{H}- & \mathrm{C}- & \underline{\overline{\mathrm{I}}}| & + & \mathrm{H}-\underline{\overline{\mathrm{O}}}-\mathrm{H} \\ & | & & & \\ & \mathrm{H} & & & \end{array}$$

Hydrogen iodide

Methyl iodide

The equations show that, if we had a substance with structure I, reaction with hydrogen iodide (HI) would produce a substance, ethyl iodide, with the formula C_2H_5I; if we had a substance with structure II, treatment with HI would result in the production of two molecules of methyl iodide, CH_3I, for every molecule of starting material. When the actual substances are treated with hydrogen iodide, ethyl alcohol reacts to give ethyl iodide, and dimethyl ether gives methyl iodide. The formation of ethyl iodide from ethyl alcohol, and of methyl iodide from dimethyl ether, can be explained only if the true molecular structure of ethyl alcohol is that depicted in structure I, and if the molecular structure of dimethyl ether is that depicted in II. Our assignment of structure is therefore complete.

Isomeric substances such as ethyl alcohol and dimethyl ether, whose molecules differ in the order in which the atoms are connected, are called *structural isomers.* Thousands of examples of structural isomerism are known. A few of these will serve to illustrate the principles involved.

Aliphatic Hydrocarbons.—Let us first consider a series of organic compounds known as the *aliphatic hydrocarbons.* The simplest of these compounds is methane, the structure of which is

```
     H
     |
 H—C—H
     |
     H
```

Methane

Only one structure can be written which satisfies the octet rule, and indeed, only one compound with the formula CH_4 exists.

The next member of this series is ethane, C_2H_6, for which again only one structure, III, can be written, as the reader should verify for himself. The next member is propane, C_3H_8, for which the only possible structure is IV.

```
     H   H              H   H   H
     |   |              |   |   |
 H—C—C—H          H—C—C—C—H
     |   |              |   |   |
     H   H              H   H   H
```

III. Ethane
Boiling point−88.6°

IV. Propane
Boiling point−42.2°

However, when a fourth carbon atom is added to the molecule to give C_4H_{10}, there are two ways of organizing the atoms. They are illustrated in V and VI. And indeed, two substances with the formula C_4H_{10} are known: *normal* butane and *iso*-butane.

```
     H   H   H   H              H       H       H
     |   |   |   |              |       |       |
 H—C—C—C—C—H            H—C———C———C—H
     |   |   |   |              |       |       |
     H   H   H   H              H       |       H
                                        |
                                    H—C—H
                                        |
                                        H
```

V. *Normal* butane
Boiling point−0.5°

VI. *Iso*-butane
Boiling point−12°

As the number of atoms in the molecular formula increases, the number of possible arrangements becomes very large indeed. Thus, while there are only two structural isomers for the hydrocarbon C_4H_{10}, both of which are known, one can write thirty-five different structural isomers for the hydrocarbon C_9H_{20}, and 62,491,178,805,831 isomers for $C_{40}H_{82}$. In agreement with the theory, thirty-five different substances with the formula C_9H_{20} have actually been prepared and identified. We must confess, however, that the theoretical number of isomers has not yet been prepared and identified for $C_{40}H_{82}$.

It must be emphasized that throughout this section we have been concerned solely with the *order* in which the atoms are linked. A structural formula such as V is intended to show only the organization of the atoms within the molecule, i.e., which atom is bonded to which; it is *not* intended to pictorialize the three-dimensional molecule. In this representation the angles between the bonds may be drawn in any arbitrary manner. For example, we must not imply from V that the carbon atoms in *normal* butane are arranged in a straight line, but merely that they are connected in the order indicated. Structures such as VII, in which the angles between the carbon atoms have been drawn differently, do not represent additional structural isomers because the order in which the atoms are linked is still the same. Structural formulas V and VII represent the same molecule.

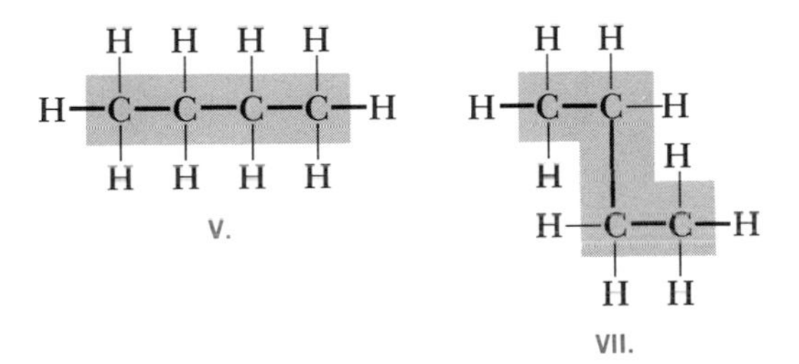

V. VII.

Physical methods of structure assignment

In the foregoing section, structural assignments were made solely on the basis of chemical reactions. In addition to the chemical methods, there are also a host of physical methods that are capable of providing information about the structure of molecules.

Many of the physical methods are spectroscopic, that is, they involve the absorption or emission of energy in the form of radiation by the molecules. Recall that, in the case of the hydrogen atom, the spectrum provided information about the energy levels of the atom (Ch. 9). The same kind of information can be obtained for molecules by a study of their absorption and emission of radiation over a wide range of the electromagnetic spectrum. Experiments of this sort provide information about molecular structure because the energy levels are characteristic of the molecular structure.

Absorption of energy in the *visible* and *ultraviolet* region results in the excitation of *electrons* in the molecule to higher energy levels. It will be recalled that for the hydrogen atom, the spectrum consists of sharp lines. For molecules, on the other hand, excitation of an electron to higher energy levels results in a spectrum consisting of broad lines or *bands* of characteristic width and shape. For instance, the absorption band responsible for the purple color of potassium permanganate in solution is shown in Fig. 14-1. This band has some highly distinctive features: two maxima, a minimum, and two "shoulders"—all at well-defined wavelengths. Permanganate ion is probably the only substance that absorbs light in precisely this way. Thus, if we measure the spectrum of an unknown sample and find the identical features, we can be reasonably certain that permanganate ion is present.

For substances with complex molecules, the absorption spectrum in the visible and ultraviolet region commonly consists of a progression of bands which, in total, are characteristic of the given kind of molecules. However, many of the bands will not be characteristic of the entire molecule but only of specific parts of it. Thus, a carbon-oxygen double bond will produce a distinctive band in a certain narrow range of wavelengths, N═N will produce a different distinctive band in a different range of wavelengths, and so on. When these distinctive absorption bands appear in the spectrum of an unknown substance, we can be reasonably sure that the corresponding bonds or groups of atoms are present in the unknown molecules.

In the *infrared* region of the spectrum, absorption of radiation leads to an increase in the energy of *vibration* of the atoms in the molecule. The infrared spectrum is particularly useful for structure assignment, because most infrared absorption bands can be assigned to specific bonds. For example, ethyl alcohol shows infrared absorption at a frequency that is characteristic of the frequency at which the O—H bond stretches. Dimethyl ether does not absorb at this frequency, showing that the molecules do not have an O—H bond, in agreement with the chemical evidence.

If spectra are studied at still longer wavelengths (far infrared and

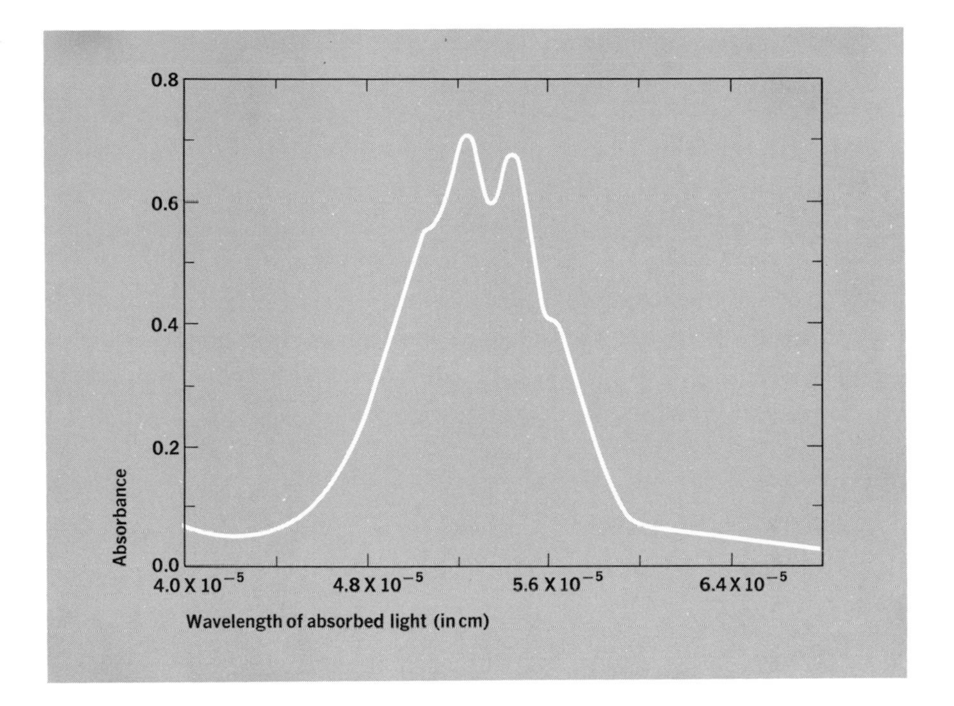

Fig. 14-1
Absorption spectrum of a dilute aqueous solution of potassium permanganate in the visible region. (The absorbance (plotted along the *Y*-axis) is a measure of the amount of light of the given wavelength that is absorbed by permanganate ion in the solution. [Courtesy of James Guzinski.]

longer), the absorptions become characteristic of rotations of, or within, the molecule. In favorable cases, such spectra will disclose the precise geometry of the molecules. Specific information about the location of atoms in a molecule, or of ions in a crystal, can be derived also from measurements of the diffraction of beams of electrons or X rays. By diffraction we mean the change in direction of parts of the beam as the result of interactions with either the nuclei or the electrons of the atoms.

Nuclear magnetic resonance is another method—a very powerful method—for determining molecular structure. One observes the behavior of certain atomic nuclei, such as those of hydrogen, fluorine or phosphorous, when a compound containing these atoms is placed in a strong magnetic field. Since the behavior of those nuclei (more technically, the *resonance absorption* of those nuclei) depends on the nature of the neighboring atoms in the molecule, information about the structure of the molecule can be obtained. For example, nuclei of hydrogen atoms in liquid ethyl alcohol show resonance absorption at three different frequencies, characteristic of the hydrogen atoms of the CH_3 group, CH_2 group, and OH group.

```
   H     H
   |     |
H—C —— C —— O—H
   |     |
   H     H
```

Ethyl alcohol

Mass spectrometry is another powerful tool for the study of molecular structure. When a molecule is struck by a fast moving electron it may undergo fragmentation. By studying the fragments that are formed, it is possible to deduce considerable information about the structure of the parent molecule.

The tetrahedral carbon atom

In the preceding sections we studied the organization of the atoms within the molecule without considering the relationship that the atoms might bear to one another in three-dimensional space. That this relationship is important, however, was recognized almost simultaneously by Louis Pasteur in France and Friedrich Augustus Kekulé von Stradonitz in Germany. Kekulé, writing in 1859, pointed out that the incompleteness of the structural formula can be avoided if, instead of arranging the four bonds of the carbon atom in a plane, we place them so that they run out and end in the corners of a tetrahedron. In 1860, Louis Pasteur, at that time interested in a phenomenon known as optical activity, suggested an explanation that involved the three-dimen-

Friederich Augustus Kekulé:
1829–1896
Originally a student of architecture, Kekulé was responsible for many of the early theories of valence and molecular structure. [Photograph courtesy Culver Pictures.]

sional arrangement of the atoms in those molecules that exhibit optical activity.

These early speculations were developed into a coherent theory of molecular structure by two chemists working quite independently, one in Holland and the other in France. In 1874, J. H. van't Hoff and J. A. LeBel both published articles to explain the phenomenon of optical activity. This phenomenon will be considered later. What concerns us at this time is that their conclusions reached far beyond the field of optical activity and applied to a wide variety of molecules.

In accord with Kekulé's earlier suggestion, van't Hoff and LeBel postulated that the four covalent bonds which the carbon atom is capable of forming are directed toward the corners of a regular tetrahedron, a regular geometrical solid shown in Fig. 14-2. The carbon

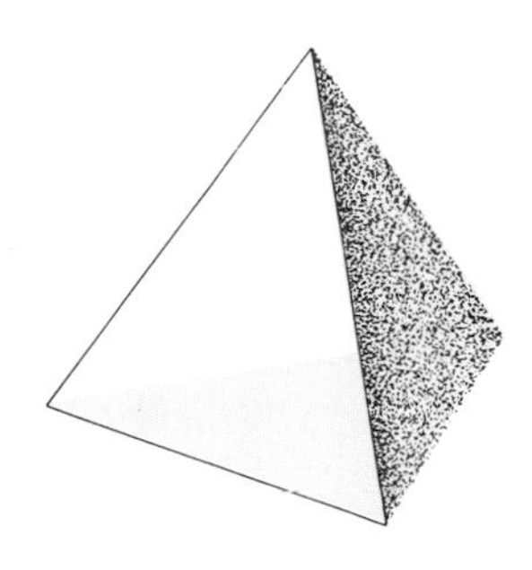

Fig. 14-2
A regular tetrahedron.

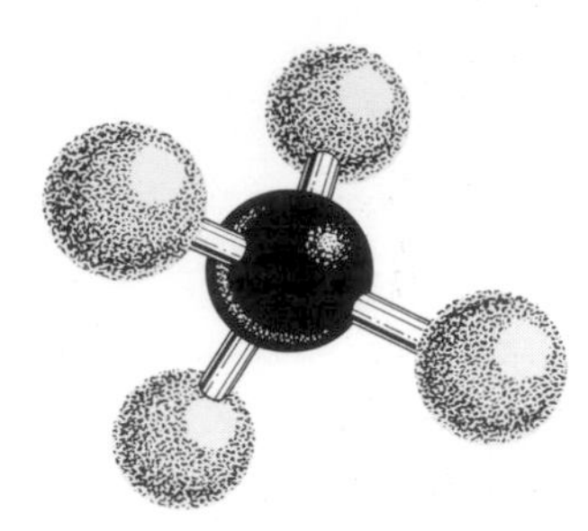

Fig. 14-3
Model of the tetrahedral methane molecule.

atom is located at the center of the tetrahedron, and the four atoms to which it is bonded are located at the corners. This arrangement is illustrated by Fig. 14-3, in which the methane molecule is used as an example.

Directed bonds and localized bonding orbitals

Can the localized bonding orbitals described in the preceding chapter account for the geometric properties of such molecules as methane? If we assume that the bonds are formed by the merging or overlap of half-filled atomic orbitals, it would be reasonable to expect that the geometry of the bonds is, to some extent, determined by the geometry of the atomic orbitals.

The only atomic orbitals of which we have any precise knowledge are those of the hydrogen atom. As shown on p. 186, an s orbital is spherical in shape and exhibits no preferential bond direction. On the other hand, a p orbital is concentrated along an axis passing through the nucleus, and the electron density extends symmetrically on both sides of the nucleus. (See Fig. 14-4a.) A p orbital can overlap best with the orbital of another atom if the other atom approaches along that axis. This effect is illustrated in Fig. 14-4b for the overlap of two p orbitals. We therefore expect bonds involving p orbitals to have strong directional properties.

Let us apply this concept to a specific case, namely the bonding of the hydrogen atoms to the phosphorus atom in the PH_3 (phosphine) molecule. The electronic configuration of a free phosphorus atom is shown below. It can be seen that the phosphorus atom has three half-filled $3p$ orbitals. Since the p orbitals in a given shell are mutually

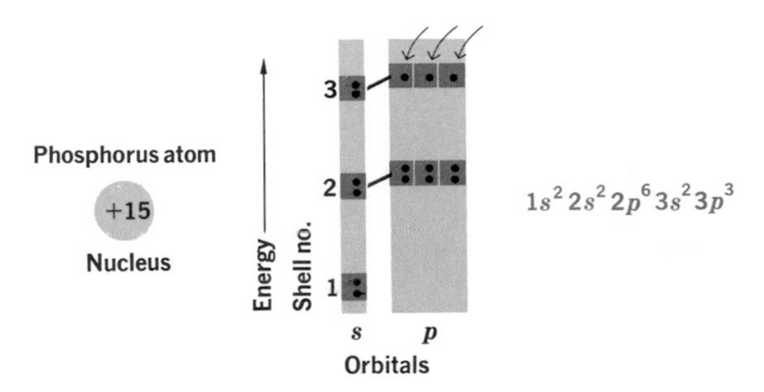

perpendicular (see, for example, Fig. 14-5), we would expect that any bonding orbitals formed from them would likewise be perpendicular. Thus we would expect the geometry of the phosphine molecule to

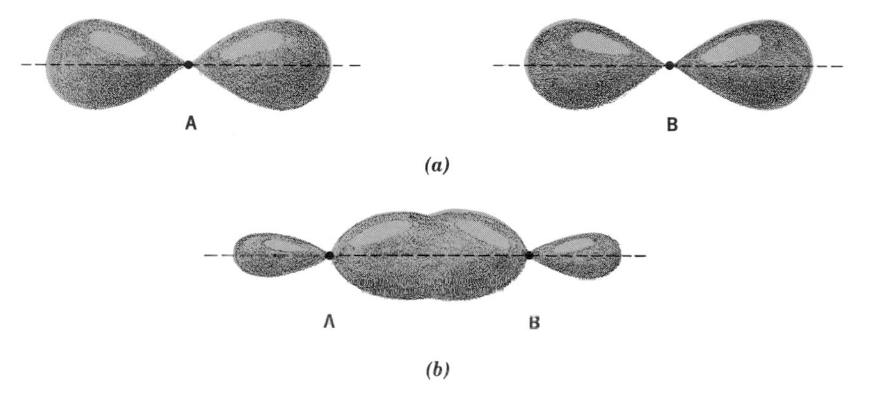

Fig. 14-4
Direction of best overlap for the p orbitals of two atoms A and B. (a) Separated atoms. (b) Localized bonding orbital.

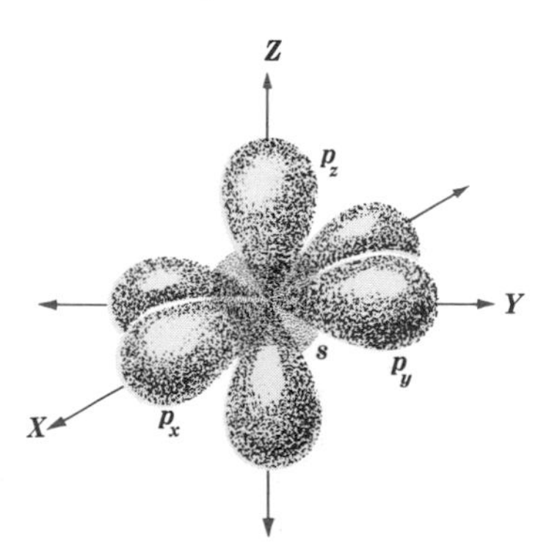

Fig. 14-5
The three p orbitals in a given shell of any free atom are mutually perpendicular. Shown here are the three $2p$ orbitals of the hydrogen atom.

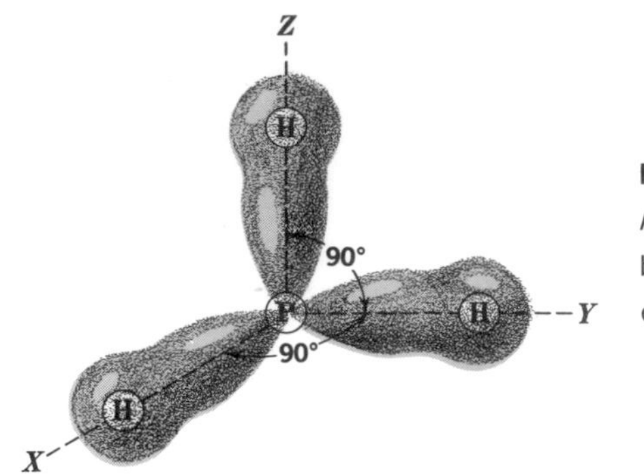

Fig. 14-6
A plausible picture of the localized bonding orbitals in the PH_3 molecule.

resemble that shown in Fig. 14-6, with three P—H bonds at angles of 90°, generated by the overlap of the three *3p* orbitals of the phosphorus atom with the 1*s* orbitals of three hydrogen atoms. In fact, a variety of experiments has confirmed this prediction of the molecular geometry.

Hybrid Orbitals.—In the case of carbon compounds, this simple picture must be modified. The electron configuration of the stable carbon atom is, $1s^2 2s^2 2p^2$, with two half-filled *2p* orbitals. However, in Chapter 13 it was pointed out that this configuration cannot account for the fact that the carbon atom ordinarily forms *four* covalent bonds. The dilemma was then resolved by introducing the concept of *bond hybridization.* If we take the electron distributions represented by the 2*s* and the three 2*p* orbitals of the hydrogen atom and mix them mathematically, we can obtain four new orbitals that are equivalent to each other. The four orbitals produced by this operation are called sp^3 hybrid orbitals. It is gratifying to discover that these four equivalent orbitals are directed toward the corners of a regular tetrahedron, in agreement with the well-established tetrahedral bond directions of the carbon atom. The sp^3 orbitals thus envisioned for the carbon atom are shown in Fig. 14-7, and the corresponding tetrahedral bonding orbitals in the methane molecule are shown in Fig. 14-8.

The NH_3 (ammonia) molecule presents a similar problem in molecular geometry. The electron configuration of the nitrogen atom is $1s^2 2s^2 2p^3$, which leads us to expect that the nitrogen atom will form three mutually perpendicular bonds involving the 2*p* orbitals, analogous to the bonds in PH_3. However, when the bond angles in the ammonia molecule were measured, they were found to be 107.3° rather than 90°. This is so near the tetrahedral angle of 109°28′, that it strongly suggests that the bonding in ammonia is similar to that in methane. As shown below, we imagine the formation of four tetrahedral sp^3 hybrid orbitals, three of which form tetrahedrally directed bonds to hydrogen atoms.

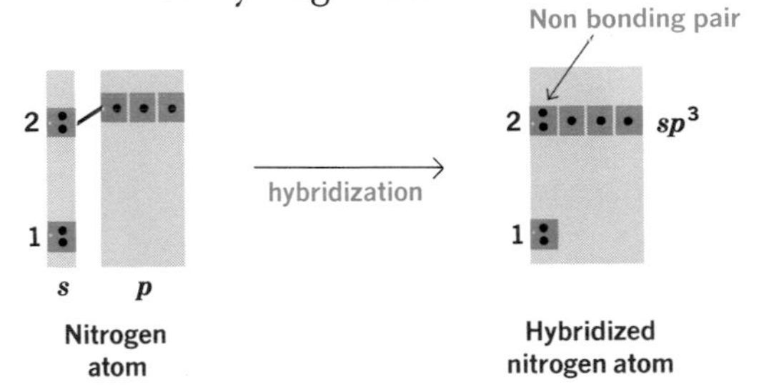

It is interesting that the water molecule is also nearly tetrahedral in shape. Presumably, the oxygen atom also forms four sp^3 hybrid orbitals, two of which form bonds to hydrogen atoms while the other

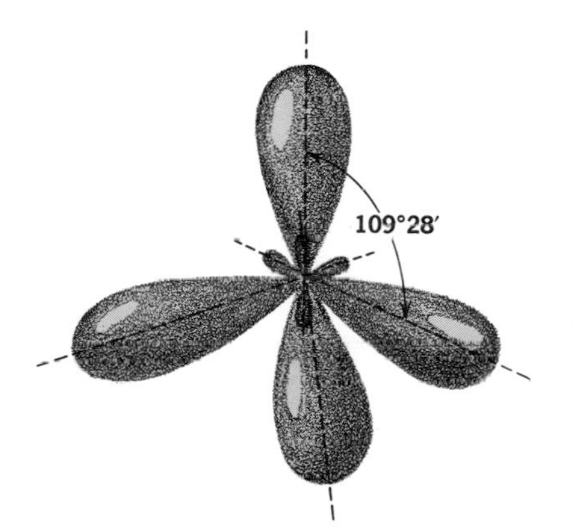

Fig. 14-7
The four sp^3 hybrid orbitals envisaged for the carbon atom.

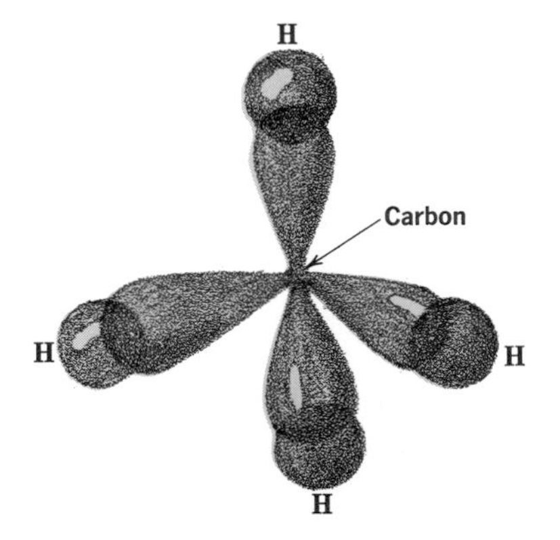

Fig. 14-8
A plausible picture of the localized bonding orbitals in the CH_4 molecule.

two are occupied by the two unshared pairs of electrons. The actual geometry is shown in Fig. 14-9.

Bond hybridization, it should be said again, is not a real physical process but merely a theoretical procedure to account for the geometrical properties of molecules. If it were possible to obtain exact mathematical solutions of the wave-mechanical equations, *molecular* orbitals could be calculated that (one hopes) would give the correct geometry without this type of *ad hoc* assumption.

MOLECULAR ARCHITECTURE OF THE ALIPHATIC HYDROCARBONS

If it is assumed that a carbon atom forming four single bonds always has a tetrahedral geometry, then the molecular architecture of the resulting molecules can be inferred directly. For a specific example, we shall examine the three-dimensional structure of some typical hydrocarbon molecules.

It has been found that in ethane (C_2H_6) the C—C bond distance is 1.54×10^{-8} cm, the C—H distances are 1.09×10^{-8} cm, and the bond angles are tetrahedral. The ethane molecule may therefore be represented by the tetrahedral ball-and-stick model shown in Fig. 14-10.

If now one joins three carbon atoms to form the propane molecule (C_3H_8), one obtains a molecule that can be depicted similarly, as in Fig. 14-11. Likewise, the two isomers of the next higher hydrocarbon, butane (C_4H_{10}), are depicted in Fig. 14-12.

It is worthwhile to contrast the three-dimensional picture of the normal butane molecule, Fig. 14-12, with its representation by means of structural formulas, as shown on page 248. In the structural formulas, the angles and distances are arbitrary: the only information conveyed is the order in which the atoms are connected. By contrast, the three-dimensional formula conveys not only that information, but also the bond angles and bond distances. Thus, we see in Fig. 14-12 that the carbon skeleton in *normal* butane is neither straight nor bent at an angle of 90° (as might be inferred from V and VII), but forms a zigzag with tetrahedral angles as required by the tetrahedral nature of the carbon bonds.

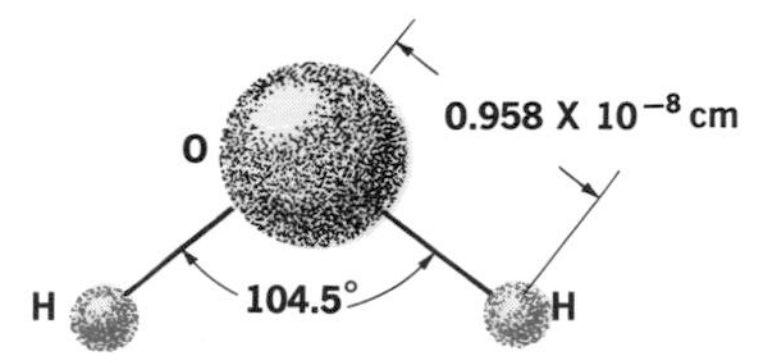

Fig. 14-9
Experimental result for the structure of the water molecule.

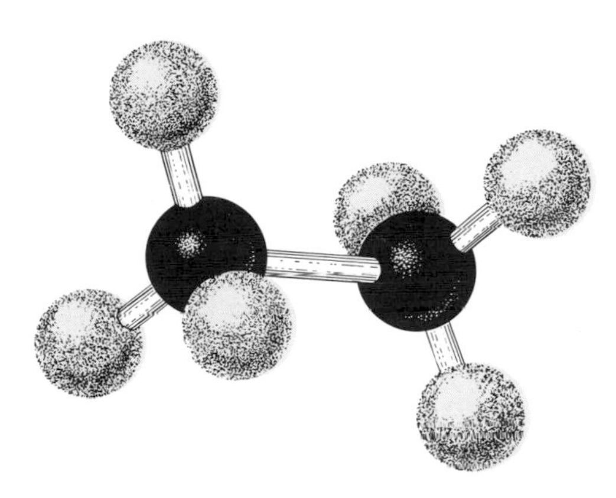

Fig. 14-10
Ball-and-stick model of the ethane molecule. The dark balls represent carbon atoms and the light balls hydrogen atoms.

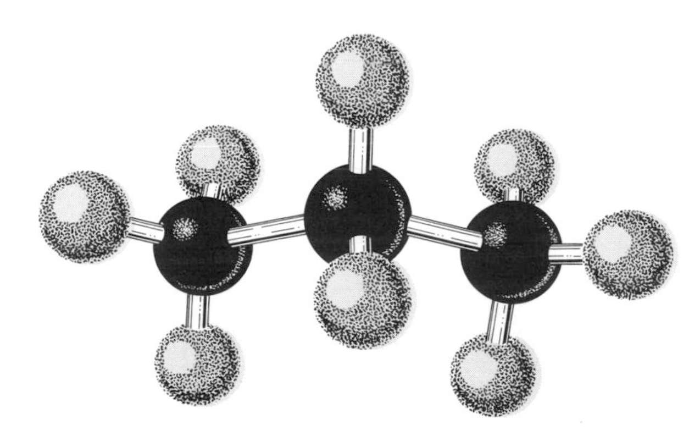

Fig. 14-11
Ball-and-stick model of the propane molecule.

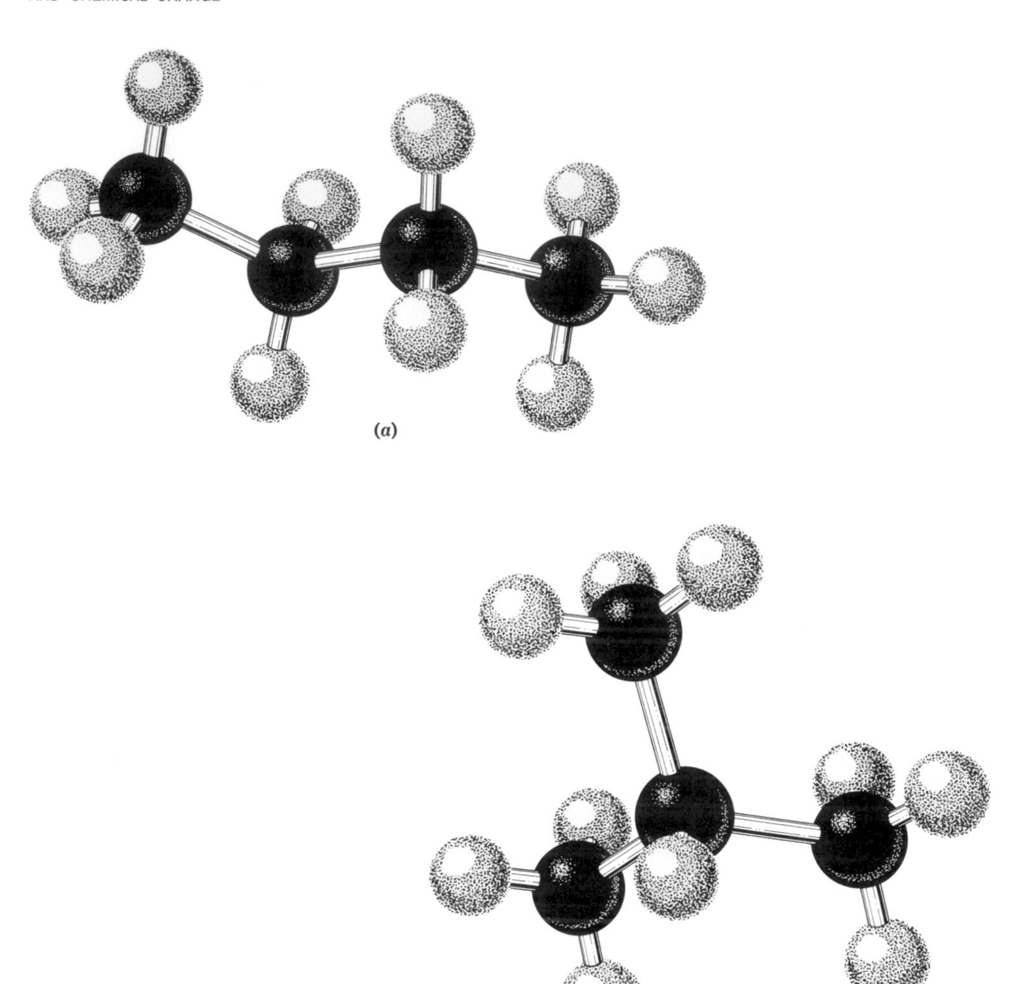

Fig. 14-12
Ball-and-stick models of the two isomers of C_4H_{10}: (a) *Normal* butane; (b) *iso*-butane.

Geometrical isomerism

A second type of isomerism that was quite puzzling for a long time is exemplified by the existence of the two substances, maleic acid and

Jacobus Hendricus Van't Hoff: 1852–1911
The Dutch physical chemist responsible (along with the Frenchman J. A. Le Bel) for the concept of the asymmetric carbon atom. [Photograph courtesy Brown Brothers.]

fumaric acid. Both acids have the molecular formula $C_4H_4O_4$, and the arrangement of the atoms is the same in both compounds, that is, the two compounds are *not* structural isomers. Yet they are clearly different compounds: maleic acid melts at 135°C, and fumaric acid at 287°C. Maleic acid is readily converted by heating to a new substance, called maleic anhydride ($C_4H_2O_3$), which is formed by the loss of water; fumaric acid is converted to the anhydride only with difficulty. Since the two substances are clearly different yet are not structural isomers, wherein lies the difference? Van't Hoff suggested that the difference was due to the spatial arrangement of the atoms in the two molecules.

To understand van't Hoff's theory of geometrical isomerism, it is necessary first to consider the rotation of two groups of atoms about the covalent bond joining them. It has been found that two groups of atoms, when joined by a *single* bond, are quite free to rotate with respect to one another. Thus, in ethane,

```
H         H
 \       /
H—C-----C—H
 /       \
H         H
```

each group can rotate about the axis of the C—C covalent bond, just as

Fig. 14-13
Free rotation about the C—C bond axis in ethane.

two wheels may rotate independently on a common axle (see Fig. 14-13). If, however, the two carbon atoms are joined by two covalent bonds (a double bond), as is the case in ethylene (C_2H_4), this rotation is no longer possible, and the structure is a relatively rigid one. To pursue the analogy of the two wheels, if the wheels were joined by two separate axles as shown in Fig. 14-14, they would no longer be free to rotate with respect to each other.

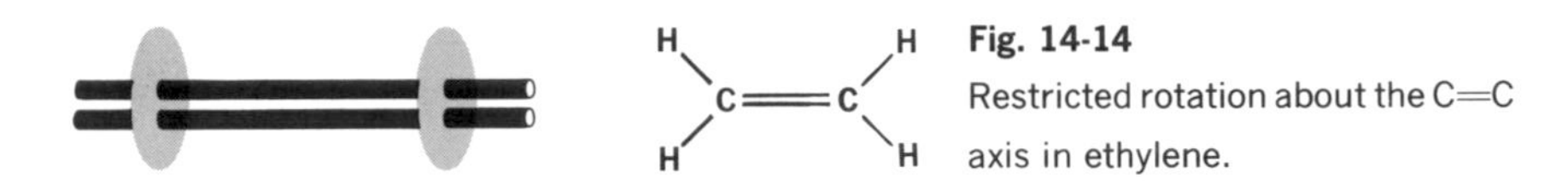

Fig. 14-14
Restricted rotation about the C═C axis in ethylene.

Since maleic and fumaric acid both contain a double bond, it was van't Hoff's hypothesis that the two molecules differed in the way in which the other atoms in the molecule were arranged about this rigid spine. He proposed the following structures for the two acids:

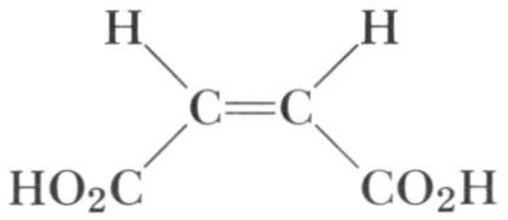

```
H         H              H         CO2H
  \     /                  \     /
   C═C                      C═C
  /     \                  /     \
HO2C      CO2H           HO2C      H
```

VIII. Maleic acid

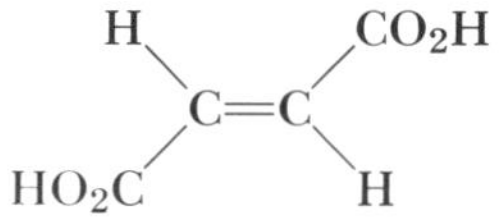

IX. Fumaric acid

The two groups joined by the double bond are not free to rotate with respect to each other, so that the groups retain their relative positions in space, and the two molecules are distinct. In the case of maleic acid, the two hydrogen atoms are on the same side of the double bond (*cis*), while in fumaric acid they are on opposite sides (*trans*). Many other examples of this type of geometrical isomerism about a double bond are known.

The freedom of rotation about single bonds of course makes it impossible for isomerism of this type to arise in analogous compounds containing a single bond instead of the double bond. For example, it is possible to write two formulas for succinic acid, $C_4H_6O_4$, which are analogous to those shown above for maleic acid and fumaric acid:

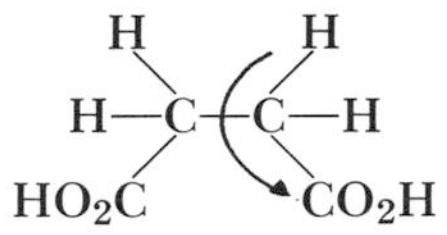

```
  H       H                      H      CO2H
   \     /                        \     /
 H—C—C—H          ⟶          H—C—C—H
   /     \                        /     \
HO2C     CO2H                 HO2C      H
```

X. (CO_2H groups on same side) 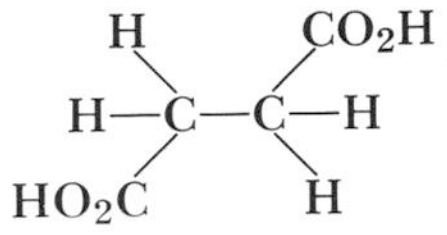 XI. (CO_2H groups on opposite sides)

In fact, only one substance with the structural formula of succinic acid is known. This is readily understood because rotation about the single bond rapidly interconverts structures X and XI. Thus, an actual sample of succinic acid consists of molecules with all possible relative orientations of the CO_2H groups.

The asymmetric carbon atom

There is yet another type of isomerism that arises out of the geometrical character of the tetrahedral carbon atom. When the four atoms or groups attached to the carbon atom are all different, the molecule becomes asymmetric. By this is meant that the molecule and its mirror image are no longer identical, and therefore two nonsuperimposable forms can exist. Consider, for example, the substance, alanine, which contains a carbon atom (C*) having four different groups attached to it: CH_3, NH_2,

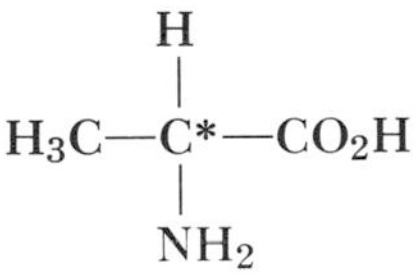

```
        H
        |
H3C—C*—CO2H
        |
       NH2
```

XII. Alanine

H, and CO_2H. Two different structures exist for this molecule, one of which is the mirror image of the other. These structures are shown in Fig. 14-15. The two isomers of alanine differ from each other in the same way as does the right hand from the left hand. How does this

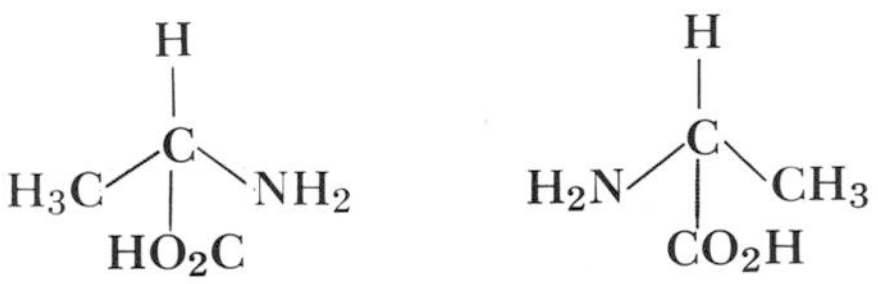

```
      H                    H
      |                    |
      C                    C
    /  | \               /  | \
 H3C   |  NH2         H2N   |  CH3
     HO2C                 CO2H
```

Alanine   Mirror image

Fig. 14-15
The two forms of alanine.

difference manifest itself? On gross examination, the two isomers appear to be identical. For example, they have the same density and the same solubility in water, and they react at equal rates with symmetrical molecules. However, a difference does manifest itself when the two isomers react with asymmetric molecules, or when they interact with polarized light.

A beam of light is said to be polarized if the wave motion associated with the light beam is restricted to a single plane. When such a beam passes through a solution containing one form of an asymmetric molecule, the plane of polarization is rotated in a clockwise direction. When the beam passes through a solution containing the mirror-image form, the plane of polarization is rotated in a counterclockwise direction. The rotation of the plane of polarization by asymmetric molecules is referred to as *optical activity*, and the pair of related molecules are referred to as *optical isomers*.

Optical isomerism is extremely important in the chemistry of living things. Many of the substances that are essential to the diet are optically active. If the organism is fed one of the isomers, it will grow and multiply, whereas it cannot utilize the mirror image at all.

Suggestions for further reading

Drexhage, K. H., "Monomolecular Layers and Light," *Scientific American*, March 1970, p. 108.

Fraser, R. D. B., "Keratins," *Scientific American*, August 1969, p. 86.

Frazer, A. H., "High-Temperature Plastics," *Scientific American*, July 1969, p. 96.

Kendrew, J. C., "The Three-dimensional Structure of a Protein Molecule," *Scientific American*, Dec. 1961, p. 96.

Lambert, J. B., "The Shapes of Organic Molecules," *Scientific American*, *222*, January 1970, p. 58.

Orchin, M., "Determining the Number of Isomers From a Structural Formula," *Chemistry*, *42*, May 1969, p. 8.

Petersen, Q. R., "Some Reflections on the Use and Abuse of Molecular Models," *Journal of Chemical Education*, *47*, January 1970, p. 24.

Robertson, D. H., and Reed, D. I., "Mass Spectrometry," *Chemistry, 42,* June 1969, p. 7.

Sullenger, D. B., and Kennard, C. H. L., "Boron Crystals," *Scientific American,* July 1966, p. 96.

Wagner, J J., "Nuclear Magnetic Resonance Spectroscopy," *Chemistry, 43,* March 1970, p. 13.

Weisskopf, V. F., "The Three Spectroscopies," *Scientific American, 218,* Jan. 1968, p. 15.

Wunderlich, B., "The Solid State of Polyethylene," *Scientific American,* November 1964, p. 80.

Questions

1. Define or explain, and give an example of:
(*a*) Structural isomers.
(*b*) Geometrical isomers.
(*c*) Optical isomers.
(*d*) Localized bonding orbital.

2. For each of the following molecular formulas, write at least two different structural formulas, using only those combinations of single, double and triple bonds that satisfy the octet rule:
(*a*) C_3H_8O; (*b*) C_3H_6; (*c*) $C_2H_4Cl_2$; (*d*) C_2H_5N;
(*e*) BCH_8N with a B—N bond; (*f*) H_3NO
(Hint: In some cases it will be necessary to consider structural formulas that involve rings of atoms.)

3. Three different substances are known with the molecular formula $C_2H_2Cl_2$. Write possible structural formulas that satisfy the octet rule.

4. Three different substances are known with the molecular formula C_5H_{12}. Write possible structural formulas that satisfy the octet rule.

5. One of the arguments against a *square* configuration of the carbon atom

is that only one substance with the formula CH_2Cl_2 exists. Explain.

6. The known substances, *cyclo*propane and *cyclo*butane, consists of molecules with the following structural formulas:

```
    H   H
    |   |
H—C—C—H
    |   |
H—C—C—H
    |   |
    H   H
```

*Cyclo*butane

```
    H   H
    |   |
H—C—C—H
     \ /
      C
     / \
   H     H
```

*Cyclo*propane

(*a*) Are the structural formulas consistent with the octet rule?
(*b*) Can the bonds formed by the carbon atoms in these molecules be tetrahedral?

7. For each of the following molecular formulas, write a structural formula and predict the bond angles in that formula:
(*a*) BH_6N, with a B—N bond; (*b*) C_2H_4O, with a carbon-oxygen double bond; (*c*) CH_4O, assuming the bond angles to the oxygen atoms are the same as in the water molecule.

8. Predict the geometry of the $H-\overline{\underline{S}}-H$ (H_2S) molecule from the directional properties of the orbitals used for bond formation. Assume that bond hybridization can be neglected. The experimental value of the H—S—H bond angle is $92°20'$.

9.* Predict what might happen if two *p* orbitals, instead of overlapping and merging end-on as in Fig. 14-4, were to overlap and merge edgewise, as follows:

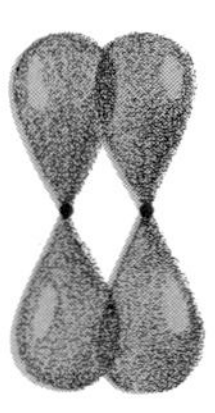

* Questions marked with an asterisk (*) are more difficult or require supplementary reading.

Would the resulting bond be a localized bond? Would rotation about the bond be free, as in Fig. 14-13, or restricted, as in Fig. 14-14?

10. Explain why a substance with the structural formula

```
     H
     |
Cl—C—NH2
     |
     Br
```

can exist as a pair of optical isomers, while

```
     H
     |
Cl—C—NH2
     |
     H
```

cannot.

11. Comment on the accuracy of the ball-and-stick model of the ethane molecule (Fig. 14-10) in light of the fact that the C—C bond distance is 1.54×10^{-8} cm and the radius of the carbon atom is approximately one half this distance.

12.* Compound **A**, whose molecular formula is C_2H_7N, has an infrared spectrum showing the presence of a carbon-carbon single bond and reacts with sodium metal with the evolution of hydrogen gas. Compound **B**, whose molecular formula is C_2H_3N, also has an infrared spectrum showing the presence of a C—C single bond but does not react with sodium metal. On the basis of this evidence:

(*a*) Deduce the structural formulas of **A** and **B**.

(*b*) Predict the structural formula for the reaction product of **A** with sodium.

Upon this principle the whole
art of making experiments in
chemistry is founded: we must always
suppose a true equality between
the principles of the body which is
examined and those which are
obtained on analysis.

ANTOINE LAURENT LAVOISIER (1743–1794)

Fifteen

CHEMICAL EQUILIBRIUM—ACIDS AND BASES

IN CHAPTER 3 we discussed the macroscopic characteristics of chemical change: the reactants that are initially present disappear, while at the same time the products, a new set of substances with quite different properties, appear. The chemical reaction is accompanied by the absorption or release of energy. The speed at which these changes take place varies widely, depending on the nature of the reaction and on the reaction conditions. In some reactions, the changes take place in a fraction of a second and, if energy is released, with explosive violence. In other reactions, the changes take place more gradually and may be observable for many minutes, or hours, or days, after the reactants are mixed. However, regardless of the speed at which the changes occur initially, sooner or later the changes become slow and eventually they become imperceptible. When the latter state is reached, the properties of the reaction mixture remain constant indefinitely, as long as the temperature, pressure, and other external conditions are kept constant. A reaction mixture that has reached a state in which no further change takes place is said to be at *equilibrium*.

Reversibility of reactions

An interesting fact about the equilibrium state is that the reaction need not be complete. Consider, for example, the reaction of hydrogen gas with iodine vapor to produce hydrogen iodide gas.

$$H_2 + I_2 \longrightarrow 2\,HI \qquad (1)$$

This reaction is very slow at room temperature, and progress can be detected only if many days are allowed to elapse between measurements. However, the reaction speeds up considerably as the temperature is raised—the reasons will be considered in the next chapter—and at temperatures above 300 °C the progress to equilibrium can be followed conveniently. It is found that when equilibrium is reached, the reaction mixture contains appreciable quantities of unreacted hydrogen and iodine.

Reactions that are incomplete at equilibrium can be reversed easily. Consider, for example, reaction (2), which is the exact reverse of reaction (1).

$$2\,HI \longrightarrow H_2 + I_2 \qquad (2)$$

When pure hydrogen iodide gas is heated above 300 °C, the formation of hydrogen gas and iodine, which is visible as a violet colored vapor, is soon evident. However, reaction (2) also fails to go to completion and leads to an equilibrium state in which all three substances—H_2, I_2, and unreacted HI— are present. Moreover, the composition of the equilibrium state resulting from (2) is identical with that resulting from (1), provided that the same numbers of hydrogen and iodine atoms and the same reaction temperature and volume are used. In other words, under comparable conditions the final equilibrium composition will be the same, regardless of the direction in which the reaction is carried out.

Reactions that can be carried out in either direction are said to be *reversible*. It is convenient to represent reversible reactions, such as (1) and (2), by using double arrows, as shown in (3).

$$H_2 + I_2 \rightleftharpoons 2\,HI \qquad (3)$$

In reversible reactions, the terms *reactants* and *products* are ambiguous unless the direction of the reaction is specified. However, it is customary to use a convention in which the reactants are the substances appearing on the left side of the equation and the products are those appearing on the right side. If the products predominate at equilibrium, we say that the equilibrium lies on the right and the reaction has gone to completion. If the reactants predominate, we say that the equilibrium lies on the left. The system at equilibrium is referred to as an "equilibrium mixture."

In theoretical discussions of chemical equilibrium, one makes no fundamental distinction between reactions that are incomplete at equilibrium and those that appear to have gone to completion. The latter are simply equilibria that lie very far on the right. This theory asserts that *some reactant is present at equilibrium in all reactions,* even when the amount is too small to be detected. Thus, in theory, all chemical reactions are reversible. This is an important generalization. If a reaction is known to be reversible, no matter how slightly, there is hope that conditions might be found under which the reaction can be reversed in practice. It is easy to see that this theory greatly enlarges the scope of chemistry, because any reaction in which a substance participates as a reactant can, if reversed, become a method of synthesis for that substance.

Energy and entropy

There are two factors which determine the degree of completion of a chemical reaction when equilibrium is reached. These are the energy factor and the entropy factor. The energy factor is already familiar: processes tend to take place spontaneously when they result in the release of energy, that is, when the products have less energy than the reactants. The other factor, the *entropy* factor, involves the tendency of natural processes to proceed in the direction of maximum submicroscopic randomness or disorder. For instance, nitrogen and oxygen will mix spontaneously, but no one has yet seen nitrogen and oxygen in air separate spontaneously. Or, to give an example involving a pure substance—above 0°C, liquid water is more stable than ice, even though liquid water has the higher energy. In this case, the greater stability may be attributed to the greater submicroscopic disorder of liquid water compared to that of crystalline, highly ordered, ice.

In a chemical reaction, the position of equilibrium is determined by the interplay of these two factors. The entropy factor drives the reaction towards the condition of greatest disorder, which is a condition in which reactants and products are present simultaneously. The energy factor drives the reaction towards the condition of lowest energy:

towards a state of "complete reaction" if the products have less energy, or towards a state of "no reaction" if the reactants have less energy. The relative importance of these opposing factors depends on the particular reaction, and in many cases the energy factor is overwhelmingly more important. However, the entropy factor is never entirely negligible, and therefore no chemical reaction ever goes strictly to completion.

How to shift an equilibrium

The process of changing the composition of an equilibrium mixture by appropriate adjustment of the reaction conditions is known as *shifting the equilibrium.* It is relatively easy to shift an equilibrium slightly, for example, to make the reaction of a particular substance more nearly complete. But to shift an equilibrium substantially or even to reverse its position is usually a difficult task that taxes all of the chemist's skill and ingenuity. In the following paragraphs we shall describe briefly a few of the stratagems that are useful for shifting an equilibrium.

1. If there are two or more reactants, and we increase the concentration of one of the reactants in the equilibrium mixture, the reaction of the other reactants will be more nearly complete.

To illustrate the effectiveness of this stratagem, let us quote some actual data, using again the reaction of H_2 with I_2 in the gas phase, as shown in Eq. (3). If we place 0.001 mole of H_2 and an equal number of moles of I_2 into a 1-liter flask and heat the mixture at 300 °C, we shall find that at equilibrium, 88% of each reactant has been converted to HI. Now if we increase the concentration of H_2 by admitting an additional 0.001 mole of hydrogen gas to the mixture and heat again at 300 °C, we shall find that when the new equilibrium is established, the reaction of the other reactant, iodine, is 99% complete. There is, of course, an excess of hydrogen gas remaining.

2. If we remove a product from the equilibrium mixture, we shift the equilibrium so that the reaction goes more nearly to completion.

To illustrate this method we shall use the reaction of hydrogen with nitrogen to form ammonia in the gas phase, as indicated by Eq. (4).

$$N_2 + 3\ H_2 \longrightarrow 2\ NH_3 \tag{4}$$

The progress of this reaction is very slow at room temperature but proceeds at a convenient rate above 400 °C. If we allow 0.01 mole of N_2 and 0.03 mole of H_2 to react at 450 °C in a 1-liter flask, the formation of ammonia at equilibrium is very incomplete, amounting to only 0.0002 mole, or just 1% of the completed reaction. However, ammonia can be removed very effectively from the equilibrium mixture achieved at 450 °C if we chill the mixture to just above −200 °C. At that temperature, the ammonia precipitates as a nonvolatile solid, whereas the unreacted H_2 and N_2 mixture remains gaseous. The gas phase can then be separated from the solid ammonia. The mixture is then re-equilibrated at 450 °C, and we again remove the new yield of ammonia by freezing. This cycle of operations can be repeated until virtually all of the N_2 and H_2 is converted to ammonia.

Products can also be removed by chemical rather than by physical methods. Consider, for example, a process in which solid silver chloride dissolves in water. The solubility of silver chloride in pure water is very low, but the little that does dissolve dissociates almost completely into silver ions and chloride ions (see p. 219). We can therefore represent the process as a reversible reaction.

$$AgCl\ (\text{solid}) \rightleftharpoons Ag^+ + Cl^-\ (\text{in aqueous solution}) \tag{5}$$

Equation (5) indicates that a larger amount of silver chloride could be brought into solution if either Ag^+ or Cl^- could be removed from the solution. A convenient way of removing Ag^+ is to allow it to react with ammonia.

$$Ag^+ + 2\ NH_3 \rightleftharpoons H_3N{-}Ag{-}NH_3^+ \tag{6}$$

Although the product, $H_3N{-}Ag{-}NH_3^+$, also exists in solution, it is a new chemical species in which the silver no longer functions as Ag^+, and the solubility of silver chloride increases greatly.

Reaction (6) is also reversible. If ammonia is expelled when we boil the solution, the equilibrium in Eq. (6) shifts to the left and Ag^+ is regenerated. As a result, the solubility of silver chloride decreases and crystals of solid silver chloride appear, i.e., the equilibrium in Eq. (5) shifts to the left.

3. If we raise the temperature of an equilibrium mixture, the equilibrium will shift toward that side of the chemical equation on which the substances have the higher energy.

On the molecular level, the effect of an increase in temperature is always to increase the proportion of high-energy molecules. Thus, if the energy of the products of a chemical reaction is higher than that of the reactants (i.e., if the reaction proceeds with the absorption of energy), then the ratio of products to reactants will increase with the temperature. For example, the conversion of hydrogen iodide to the elements [Eq. (2)] proceeds with the absorption of energy. When equilibrium is reached at 300 °C, the conversion to elements is 12% complete. At 500 °C, 19% of the hydrogen iodide is decomposed to the elements.

4. If we increase the pressure on an equilibrium mixture, the equilibrium will shift toward that side of the chemical equation on which the substances occupy the smaller volume.

For example, consider again the synthesis of ammonia from the elements in the gas phase, as indicated by Eq. (4). The volume occupied by two moles of gaseous ammonia is two times the gram molecular volume (p. 100), whereas that occupied by a total of one mole of nitrogen and three moles of hydrogen is four times the gram molecular volume, or twice as large. Therefore, the equilibrium shifts and favors the formation of ammonia as the pressure is increased. Such effects are important only when some of the products are gaseous under the conditions of the reaction.

5. Reactions that are virtually complete at equilibrium can often be reversed, but there are no general rules on how to do it.

Reactions that go virtually to completion are almost always accompanied by the release of large quantities of energy. To reverse them, we must restore the energy; but there are constraints on how this may be done. When energy is pumped indiscriminately into a mixture of substances, all sorts of reactions become possible, and the desired reversal of the original reaction may be only an insignificant side-reaction. It is therefore necessary to find conditions under which the energy that is supplied is used selectively to bring about the desired reversal.

Physical methods of supplying energy for selective use include irradiation with light of a specific wavelength, or electrolysis at a specific voltage. For example, the reaction of sodium with chlorine can be reversed in a suitably constructed electrolytic cell (p. 214). In the chemical methods, one brings about the desired reversal by harnessing the energy released in some other reaction. Conditions are found under which the two reactions are coupled, that is, the auxiliary reaction that provides the energy cannot proceed unless the desired reaction proceeds at the same time. This coupling of two reactions is analogous to the relation established by two children sitting on opposite sides of a seesaw. One child can go down only as the other child goes up.

The conditions under which a coupled reaction can be brought about are often as follows: Suppose that each of two substances, A and B, will react separately with a common third substance, C.

$$\mathrm{A} + \mathrm{C} \rightleftarrows \text{(Products of A + C)} \qquad (7a)$$

$$\mathrm{B} + \mathrm{C} \rightleftarrows \text{(Products of B + C)} \qquad (7b)$$

Let us further suppose that B is the more powerful reagent, that is, reaction (7b) is more nearly complete at equilibrium than is reaction (7a). It is then probable that substance A can be synthesized by the addition of substance B to the *product* of the reaction of A. Let us consider a specific example.

Acetyl chloride, a colorless liquid, is a very effective reagent for water and reacts with all traces of moisture with which it comes in contact. The products are acetic acid and HCl, as shown in Eq. (8).

$$H_3C\text{—}C(=\overline{\underline{O}})\text{—}\overline{\underline{Cl}}| + HOH \longrightarrow H_3C\text{—}C(=\overline{\underline{O}})\text{—}\overline{\underline{O}}\text{—}H + HCl \qquad (8)$$

Acetyl chloride (b.p. 52°) — Acetic acid (b.p. 118°)

Phosphorous pentachloride reacts even more readily with water. The relevant reaction is shown in Eq. (9).

$$PCl_5 + HOH \longrightarrow POCl_3 + 2\ HCl \qquad (9)$$

Phosphorous pentachloride (m.p. 167° in a sealed tube) — Phosphorous oxychloride (b.p. 105°)

If phosphorous pentachloride is added to acetic acid, reaction (10) takes place and acetyl chloride is produced.

$$PCl_5 + H_3C{-}C\begin{matrix}\diagup\!\!\diagup \overline{\underline{O}} \\ \diagdown \overline{\underline{O}}{-}H\end{matrix} \longrightarrow H_3C{-}C\begin{matrix}\diagup\!\!\diagup \overline{\underline{O}} \\ \diagdown \overline{\underline{Cl}}| \end{matrix} + POCl_3 + HCl \qquad (10)$$

Note that reaction (10) incorporates reaction (9) in the sense that PCl_5 is converted to $POCl_3$; and it reverses reaction (8) in the sense that acetyl chloride is formed from acetic acid. Since acetic acid and phosphorous pentachloride are readily available, reaction (10) is a practical synthesis of acetyl chloride, a useful chemical reagent.

Dynamic balance at equilibrium

On the macroscopic level, the equilibrium state is, by definition, a static state. The composition of the equilibrium mixture remains constant indefinitely, as long as the temperature, pressure, and other external conditions are constant. But on the submicroscopic level, the equilibrium state is a dynamic steady state in which reactants and products disappear as fast as they are formed. During any given interval of time, there is some chemical reaction taking place in the forward direction. At the same time, there is an equal amount of reaction in the reverse direction, so that the net change in composition is zero. Therefore, we say that the equilibrium state is a state of dynamic balance.

This dynamic picture of chemical equilibrium is of course wholly consistent with the concepts of molecules-in-motion and submicroscopic restlessness, which have become so firmly rooted in our science. But dynamic balance is more than a picture. It is an established fact that can be demonstrated directly by means of tracer atoms. Let us consider an example.

Iodine atoms, as they occur in nature, consist of a single isotope (p. 311), whose mass is 127 atomic weight units. However, by means of nuclear reactions it is possible to obtain iodine atoms of mass 129. These atoms have the same chemical properties as those of mass 127, but they can easily be distinguished from the latter because they are radioactive and emit beta particles. Now let us suppose that a small quantity of radioactive iodine is added to a large quantity of H_2, I_2, and HI that is in equilibrium at 300 °C. We shall specify that the added amount of radioactive iodine be so small that the mixture is not displaced significantly from equilibrium. If the equilibrium were a state of dynamic balance, the reaction of H_2 with I_2 would continually take place, and its progress could be traced through the appearance of radioactive iodine atoms in the HI species. In fact, the radioactive iodine atoms will change over gradually from the I_2 species to the HI species, as expected, until their distribution between the two species is completely random.

Acids in water

There is an important class of substances, called *acids,* whose discovery can be traced back to the very beginnings of chemistry. This class includes many well-known and powerful chemicals, such as nitric acid and sulfuric acid, that are valued by chemists because they react with a wide range of substances, speed up the progress of many reactions, and dissolve many water-insoluble minerals and ores. In this section we shall discuss the behavior of acids in aqueous solution, because it provides a good example of equilibrium in a reaction that many substances can undergo.

When dissolved in water, acids will produce solutions that taste sour, turn blue litmus† red, and conduct electric current. A great variety of substances will produce these effects, yet all such substances have one property in common: the molecules that exist in solution contain at least one hydrogen atom. We therefore infer, by straightforward induction, that the peculiar chemical property that we call *acidity* is associated with the hydrogen atom. Yet this property is not shared by all hydrogen atoms—there are many compounds of hydrogen that are not noticeably acidic in aqueous solution.

Since the solution of an acid in water conducts electric current, it is probable that the acidic hydrogen atom reacts in such a way that ions are produced. In the previous discussion of ionic compounds (Ch. 12), we assumed that ions are formed by electron transfer. To interpret

† Litmus is an organic dye that is prepared from certain lichens which are mashed, treated with potassium carbonate and ammonia, and allowed to ferment. In the laboratory, it is convenient to test for acids by placing a drop of the solution on blue litmus paper, which is an adsorbent paper that has been impregnated with the dye.

acidity, we shall now make the further assumption that *ions can be formed also by the transfer of positively charged hydrogen nuclei.* Both theories postulate that ions are formed by the transfer of a simple subatomic particle; and in both theories this particle has unit charge.

If we use a notation in which the hydrogen *atom,* which consists of a hydrogen nucleus and one electron, is denoted by H, then the hydrogen *nucleus* must be denoted by H^+ and is, therefore, equivalent to a hydrogen ion. Thus, the transfer of a positively charged hydrogen nucleus can be represented as the removal of H^+ from one molecule and its addition to another molecule.

Let us apply this theory to a specific acid, hydrogen chloride, in water. The formation of ions in such a solution then takes place by the process shown in Eq. (11).

$$HCl + H_2O \rightleftarrows H_3O^+ + Cl^- \tag{11}$$

Gain of H+
Loss of H+

The transfer of H^+ from HCl to H_2O results in the formation of two ions, H_3O^+ (called *hydronium ion*), and the familiar chloride ion. That an ion with the formula H_3O^+ should be a stable entity can be understood readily in terms of accepted principles of compound formation. In the notation of Chs. 12 and 13, the hydrogen ion must be represented as

$$[H\,\square]^+$$

thus showing the presence of a vacant 1*s* orbital. By using this orbital, the hydrogen ion can form a coordinate-covalent bond to the water molecule.

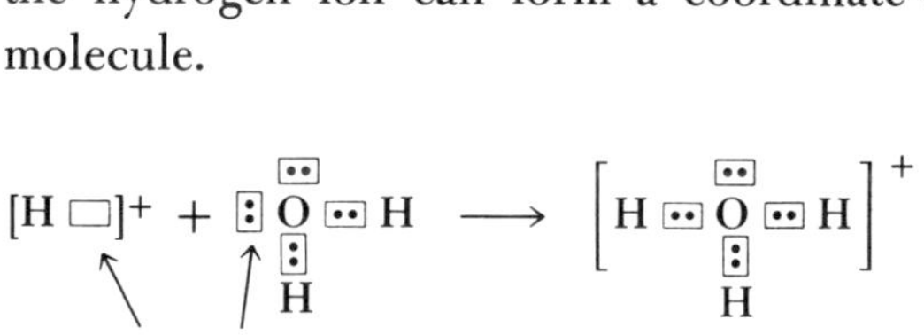

$$[H\,\square]^+ + \boxed{:}\,\overset{\boxed{\cdot\cdot}}{\underset{\boxed{:}\atop H}{O}}\,\boxed{\cdot\cdot}\,H \longrightarrow \left[H\,\boxed{\cdot\cdot}\,\overset{\boxed{\cdot\cdot}}{\underset{\boxed{:}\atop H}{O}}\,\boxed{\cdot\cdot}\,H\right]^+$$

These orbitals overlap and hydronium ion is formed.

The product, hydronium ion, is expected to be a stable unit because, as a result of electron sharing, each atom has acquired a noble-gas electron configuration.

Returning to a discussion of acids in water, we find it convenient to introduce the symbol HA to denote an acid molecule. H stands for the acidic hydrogen atom and A for the rest of the molecule. The reaction that takes place when the acid dissolves in water can then be represented by Eq. (12).

$$HA + H_2O \rightleftharpoons H_3O^+ + A^- \tag{12}$$

The position of equilibrium in (12) depends on the strength of the acid. A strong acid will readily transfer hydrogen ions to water, and the equilibrium lies far on the right. Hydrogen chloride is such an acid. At equilibrium, the reaction of HCl with water is virtually complete, and this fact is indicated in Eq. (11) where one arrow is much longer than the other. Other well-known acids whose reaction with water is nearly complete include perchloric acid ($HClO_4$), nitric acid (HNO_3), and sulfuric acid (H_2SO_4). The latter has two acidic hydrogen atoms per molecule.

$$H_2SO_4 + H_2O \rightleftharpoons H_3O^+ + HSO_4^- \tag{13a}$$

$$HSO_4^- + H_2O \rightleftharpoons H_3O^+ + SO_4^= \tag{13b}$$

Equations (13) illustrate that a molecule can have more than one acidic hydrogen atom, and that an ion (in this case, bisulfate ion, HSO_4^-) can also be an acid. Reaction (13a) goes virtually to completion, but reaction (13b) under many conditions is incomplete at equilibrium.

For most substances that are acids in water, the equilibrium in Eq. (12) lies on the left. Acetic acid is an example. At equilibrium, only a small fraction of the acetic acid molecules is converted to hydronium ion and acetate ion.

$$H_3C{-}C(=\overline{O}){-}\overline{O}{-}H + H_2O \rightleftharpoons H_3O^+ + \left[H_3C{-}C(=\overline{O}){-}\overline{O}|\right]^- \tag{14}$$

Acetic acid — Acetate ion

In the case of sucrose or cane sugar, $C_{12}H_{22}O_{11}$, the equilibrium in Eq. (12) lies so far on the left, in spite of 22 hydrogen atoms per molecule, that sugar is not normally regarded as an acid in water. And, of course, sugar solutions are sweet, not sour.

Acid-base reactions

The reaction of acids with water illustrates a very general phenomenon known as acid-base reactions. Although some acid-base reactions have been known for ages, the great generality of this phenomenon was not appreciated until the early 1900's. And as the understanding of the phenomenon improved, the concepts of "acid" and "base" underwent subtle changes and came to mean different things to different chemists. Some of the resulting confusion still lingers, but most chemists nowadays accept the definitions proposed by J. N. Brønsted, a Danish chemist, and his British contemporary, T. M. Lowry.

According to Brønsted and Lowry, an acid-base reaction is a chemical reaction in which a hydrogen ion is transferred from a donor molecule, the acid, to an acceptor molecule, the base. For example, in Eq. (14), if reaction proceeds from left to right, a hydrogen ion is transferred from the acetic acid molecule to the water molecule. Hence, acetic acid is the acid and water is the base. If the same reaction proceeds in the reverse direction, a hydrogen ion is transferred from the hydronium ion to the acetate ion. Hence, H_3O^+ is the acid, and acetate ion is the base.

Or consider the reversible reaction of sulfuric acid with chloride ion, shown in Eq. (15), which takes place when sodium chloride is dissolved in concentrated sulfuric acid.

$$\underset{\text{Base}\rightarrow}{Cl^-} + \underset{\text{Acid}\rightarrow}{H_2SO_4} \rightleftharpoons \underset{\leftarrow\text{Base}}{HSO_4^-} + \underset{\leftarrow\text{Acid}}{HCl} \tag{15}$$

When this reaction takes place in the forward direction, Cl^- is the base, and H_2SO_4 is the acid. When it takes place in the reverse direction, HSO_4^- (bisulfate ion) is the base, and HCl is the acid. These examples show that when an acid reacts with a base, the products are the acid and base for the reverse reaction.

For a substance to be a good hydrogen-ion acceptor, its molecules (or ions) must have at least one unshared electron pair. This condition is satisfied by a wide variety of substances, including all bases mentioned previously. In the following table we show the structural formulas for a few bases and for the conjugate acids. Note that the formulas for some of the bases have more than one unshared electron pair, and that all formulas satisfy the rules for stable molecule formation.

SOME ACID-BASE PAIRS

Base		Conjugate acid	
Ammonia	H—N̄—H │ H	Ammonium ion	[H │ H—N—H │ H]$^{+}$
Hydroxide ion	[|O̲̅—H]$^{-}$	Water	H—O̲̅—H
Dimethyl ether	H H │ │ H—C—O̅—C—H │ │ H H	Conjugate acid of dimethyl ether	[H H H │ │ │ H—C—O—C—H │ │ H H]$^{+}$

Many substances can react either as an acid or as a base, because the molecules have both an acidic hydrogen atom and an unshared electron pair. A familiar example is the water molecule. In Eqs. (11) to (14), water accepts a hydrogen ion and is the base. In the reaction with ammonia, shown in Eq. (16), water donates a hydrogen ion and is the acid.

$$\underset{\text{Acid}\rightarrow}{H_2O} + \underset{\text{Base}\rightarrow}{NH_3} \rightleftharpoons \underset{\leftarrow\text{Acid}}{NH_4^+} + \underset{\leftarrow\text{Base}}{OH^-} \tag{16}$$

In another reversible reaction, shown in Eq. (17), one water molecule transfers a hydrogen ion to another water molecule.

$$\underset{\text{Acid}\rightarrow}{H_2O} + \underset{\text{Base}\rightarrow}{H_2O} \rightleftharpoons \underset{\leftarrow\text{Acid}}{H_3O^+} + \underset{\leftarrow\text{Base}}{OH^-} \tag{17}$$

In this reaction, water functions both as an acid and as a base. Eq. (17) is similar to (16) except that H_2O is used as the base instead of NH_3.

The position of equilibrium in an acid-base reaction can be predicted, at least qualitatively, from a knowledge of acid or base strength. A base is strong if its molecules form a strong bond to the added hydrogen ion; it is weak if the molecules form only a weak bond to the added hydrogen ion. Conversely, if the bond to the hydrogen ion is strong, the hydrogen ion cannot be easily transferred to another molecule, and the acid is weak. Thus, there is an inverse relationship between acid and base strength: strong bases form weak conjugate acids, and weak bases form strong conjugate acids.

In the following table we have arranged some common acids and bases in order of decreasing acid strength or increasing conjugate base strength.

	Acid	Conjugate base	
Acid strength decreases ↓	$HClO_4$	ClO_4^-	Base strength increases ↓
	H_2SO_4	HSO_4^-	
	HCl	Cl^-	
	HNO_3	NO_3^-	
	H_3O^+	H_2O	
	HF	F^-	
	Acetic acid	Acetate ion	
	NH_4^+	NH_3	
	H_2O	OH^-	

This arrangement produces the result that the bond to the hydrogen ion becomes stronger as we read from top to bottom in the table. Hence, if an acid reacts with a base that appears below it in the table, the hydrogen ion is transferred to a more stable bond and the reaction will tend to go to completion.

For example, the reaction of H_3O^+ will tend to go to completion with F^-, acetate ion, NH_3, and OH^-, all of which appear below it in the table. On the other hand, the reaction of H_3O^+ tends to be very *in*-complete at equilibrium with NO_3^-, Cl^-, HSO_4^-, and ClO_4^-, all of which appear above it in the table.

Suggestions for further reading

King, E. L., *How Chemical Reactions Occur* (New York: W. A. Benjamin Co., Inc., 1963).

Luder, W. F., and Zuffanti, S., *The Electronic Theory of Acids and Bases* (New York: John Wiley & Sons, Inc., 1951).

Porter, G., "The Laws of Disorder," *Chemistry, 41,* May 1968, p. 23.

Van Ness, H. C., *Understanding Thermodynamics* (New York: McGraw-Hill Book Company, 1969).

Questions

1. How can one tell, experimentally, when a reaction mixture has reached a state of equilibrium?

2. How can one tell, experimentally, that the equilibrium state is a dynamic steady state in which reactants and products disappear as fast as they are formed?

3. Give an example (not necessarily from the field of chemistry) of a state of static equilibrium; of a state of dynamic equilibrium.

4. Why doesn't the water in a fishbowl evaporate if the relative humidity of the air above it is 100%?

5. What is meant by a reversible chemical reaction? Is the acid-base reaction of HCl with H_2O reversible? Is the combustion of logs in a fireplace reversible?

6. Consider the interplay of the energy factor and the entropy factor in a chemical reaction. How do these factors affect the degree of completion of the reaction at equilibrium? Why does the entropy factor drive the reaction towards a condition in which reactants and products are present simultaneously?

7. Consider the equilibrium, $N_2O_4 \rightleftharpoons 2NO_2$, which is established in the gas phase. The forward reaction ($N_2O_4 \longrightarrow 2NO_2$) is endothermic (absorbs energy). Predict how the equilibrium will shift if:
(*a*) The pressure of the equilibrium mixture is reduced.
(*b*) The temperature is raised.
(*c*) NO_2 is removed from the reaction vessel.

8. Consider the equilibrium, $2H_2 + CO \rightleftharpoons CH_4O$, which is established in the gas phase. The forward reaction ($2H_2 + CO \longrightarrow CH_4O$) is exothermic. Predict how the equilibrium will shift if:
(*a*) The pressure of the equilibrium mixture is raised.
(*b*) The hydrogen gas is added in excess.
(*c*) CH_4O is added.
(*d*) The temperature is raised.

9. Define, and give an example of, an acid-base reaction.

10. Write chemical equations for the following acid-base reactions. Label the acids and bases in each equation.
(*a*) HI + HOH
(*b*) $HNO_3 + NH_3$
(*c*) $HCl + F^-$
(*d*) $H_2O + F^-$

11. Arrange the following substances in order of increasing acid strength: HF, $HClO_4$, acetate ion, water, Cl^-.

12. Examination of electron-dot formulas will show that all of the bases in the table on p. 281 have at least one unshared pair of electrons. Can this observation be generalized to provided a more general definition of acids and bases?

Sire,
I have no need of *that* hypothesis.

PIERRE SIMON, Marquis de Laplace (1749–1827)

Sixteen

REACTIONS OF COVALENT MOLECULES

IN PREVIOUS chapters, we considered the nature of covalent bonding and the structure of molecules. In this chapter, we shall apply these concepts to a study of the reactions of covalent molecules and learn some of the more important classes of reactions. We shall consider the following questions: What is the nature of the intermediate steps involved in the reaction? What are the factors that determine the rates of chemical reactions? What are the factors that determine the nature of the final products? It is a knowledge of the answers to these questions that gives the chemist some control over the outcome of the reactions he is carrying out.

Reaction mechanisms

A reaction mechanism is a theory of the nature of the submicroscopic events whereby reactant molecules are converted to product molecules.

The chemist, on the basis of the experimental facts available to him concerning the reaction, constructs a hypothetical map of the submicroscopic path leading from reactants to products. These maps, while admittedly speculative, are of considerable value because they provide a way of predicting new reactions and of suggesting ways in which a given reaction can be carried out more conveniently. A reaction mechanism shares with any good theory the characteristic that it can be tested by experiment.

Reaction mechanisms generally assume that the formation of the product molecules from the reactant molecules proceeds through intermediate states that are unstable because the octet rule is violated. Consider, for example, the reaction of methyl iodide gas (CH_3I) with hydrogen iodide gas (HI) to produce methane (CH_4) and iodine (I_2):

$$\mathrm{H{-}\overset{\displaystyle H}{\underset{\displaystyle H}{C}}{-}\overline{\underline{I}}|} \;+\; \mathrm{|\overline{\underline{I}}{-}H} \;\longrightarrow\; \mathrm{H{-}\overset{\displaystyle H}{\underset{\displaystyle H}{C}}{-}H} \;+\; \mathrm{|\overline{\underline{I}}{-}\overline{\underline{I}}|}$$

This reaction involves the replacement in the reactant molecules of a C—I bond and an H—I bond by a C—H bond and an I—I bond. Let us speculate how this reshuffling of bonds might be accomplished. One theory we might propose involves an initial scission of the old bonds, followed by combination of the fragments in a new arrangement. This sequence of events is shown in Eqs. (1) to (4).

EQUATION (1) The C—I bond breaks to form two unstable fragments in which the octet rule is violated.

$$\mathrm{H{-}\overset{\displaystyle H}{\underset{\displaystyle H}{C}}{\vdots}\overline{\underline{I}}|} \;\longrightarrow\; \mathrm{H{-}\overset{\displaystyle H}{\underset{\displaystyle H}{C}}\boxed{\cdot}} \;+\; \mathrm{\boxed{\cdot}\overline{\underline{I}}|} \qquad (1)$$

EQUATION (2) The H—I bond breaks to form two unstable fragments in which the octet rule is also violated.

$$\mathrm{H{-}\overline{\underline{I}}|} \;\longrightarrow\; \mathrm{H\boxed{\cdot}} \;+\; \mathrm{\boxed{\cdot}\overline{\underline{I}}|} \qquad (2)$$

EQUATIONS (3) AND (4) The unstable fragments combine to form stable molecules of methane and iodine.

$$\mathrm{H{-}\overset{\displaystyle H}{\underset{\displaystyle H}{C}}\boxed{\cdot}} \;+\; \mathrm{\boxed{\cdot}H} \;\longrightarrow\; \mathrm{H{-}\overset{\displaystyle H}{\underset{\displaystyle H}{C}}{-}H} \qquad (3)$$

$$\mathrm{|\overline{\underline{I}}\boxed{\cdot}} \;+\; \mathrm{\boxed{\cdot}\overline{\underline{I}}|} \longrightarrow \mathrm{|\overline{\underline{I}}{-}\overline{\underline{I}}|} \qquad (4)$$

It should be noted that the unstable fragments could also combine to reform the original molecules.

In the theory represented by Eqs. (1)–(4), the reactant molecules first had to break into reactive fragments before the products could be formed. An alternative theory we might consider is that the new bonds are being formed while the old bonds are still in the process of being broken. This mechanism is shown in Eqs. (5) and (6).

EQUATION (5) The molecules of methyl iodide and hydrogen iodide combine to form an unstable intermediate molecule in which the bonds indicated by dotted lines are different from ordinary shared electron-pair bonds.

$$\begin{array}{c} H \\ | \\ H-C-I \\ | \\ H \end{array} \quad + \quad H-I \quad \longrightarrow \quad \begin{array}{ccc} H & & I \\ & \diagdown & \\ H- & C & \quad I \\ & \diagup & \\ H & & H \end{array} \qquad (5)$$

EQUATION (6) The intermediate molecule breaks up to form methane and iodine.

$$\begin{array}{ccc} H & & I \\ & \diagdown & \\ H- & C & \quad I \\ & \diagup & \\ H & & H \end{array} \quad \longrightarrow \quad \begin{array}{c} H \\ | \\ H-C-H \\ | \\ H \end{array} \quad + \quad I \; I \qquad (6)$$

Which, if either, of these two mechanisms is the correct one can be decided only by experiment. Moreover, the particular mechanism that operates may change as the reaction conditions are changed. The important thing to notice about both mechanisms is that the atoms must momentarily exist in unstable combinations before the product molecules are formed.

THE ACTIVATED COMPLEX

In the conversion of reactants to products, the intermediate state with the highest energy is called the *activated complex*, and the required amount of energy is called the *activation energy*. Figure 16-1 shows how the energy requirements in a chemical reaction can be represented diagrammatically. The high energy of the activated complex naturally

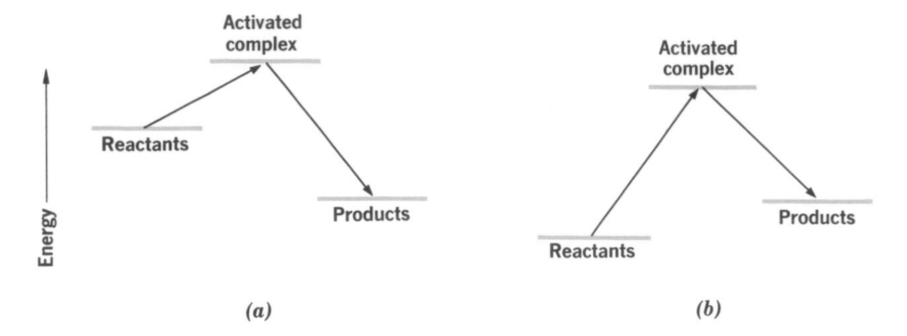

Fig. 16-1
Energy-level diagrams for chemical reactions. (a) For an ***exothermic*** reaction, which proceeds with release of energy. The energy level for the products is lower than that for the reactants. (b) For an ***endothermic*** reaction, which proceeds with absorption of energy.

raises the question: From where does the activation energy come? While the average energy of the molecules may be insufficient to form the activated complex, there is in any collection of molecules always a small fraction with energies well above the average (Ch. 5, pp. 77–80). It is these molecules that, at any given instant, are capable of undergoing reaction.

Selectivity of reagents and the principle of least structural change

Examination of a large number of chemical reactions that involve the making and breaking of covalent bonds leads to a generalization known as the *principle of least structural change.* According to this principle, a reagent will attack only one bond or, at most, a small number of bonds, in the original molecule, thus producing a new molecule in which much of the original structure is still intact. For example, the reaction that ensues when ethyl bromide is added to a solution of the ionic compound, sodium hydroxide, is represented by Eq. (7).

$$\begin{array}{ccccccccc} & \text{H} & \text{H} & & & & & \text{H} & \text{H} & & & \\ & | & | & & & & & | & | & & & \\ \text{H}- & \text{C}- & \text{C}- & \overline{\underline{\text{Br}}}| & + & {}^{\ominus}|\overline{\underline{\text{O}}}-\text{H} & \longrightarrow & \text{H}-\text{C}- & \text{C}- & \overline{\underline{\text{O}}}-\text{H} & + & |\overline{\underline{\text{Br}}}|^{\ominus} \\ & | & | & & & & & | & | & & & \\ & \text{H} & \text{H} & & & & & \text{H} & \text{H} & & & \end{array} \qquad (7)$$

Ethyl bromide | Hydroxide ion from sodium hydroxide | Ethyl alcohol | Bromide ion forms sodium bromide

Comparison of the structural formulas of the product ethyl alcohol with that of ethyl bromide reveals that the

$$\begin{array}{ccc} & \text{H} & \text{H} \\ & | & | \\ \text{H}- & \text{C}- & \text{C} \\ & | & | \\ & \text{H} & \text{H} \end{array}$$

part of the molecule is unchanged. The only bond that has been attacked is the C—Br bond, and the only structural change is the replacement of bromine by hydroxyl.

Sometimes the reaction zone consists of a small number of adjacent bonds or atoms. For example, consider the reaction of ethylene diiodide with iodide ion, which is shown in Eq. (8).

$$\begin{array}{cccccccc} & |\overline{\text{I}}| & |\overline{\text{I}}| & & & & & & \\ & | & | & & & & & & \\ \text{H}- & \text{C}- & \text{C}-\text{H} & + & |\overline{\underline{\text{I}}}|^{\ominus} & \longrightarrow & \text{H}-\text{C}=\text{C}-\text{H} & + & I_3^- \\ & | & | & & & & \;\;|\quad\;\; | & & \\ & \text{H} & \text{H} & & & & \;\;\text{H}\quad\;\text{H} & & \end{array} \qquad (8)$$

Reaction zone

Ethylene diiodide | Iodide ion | Ethylene | Triiodide ion

In this reaction, two C—I bonds are broken as a result of the attack by iodide ion, and a carbon-carbon double bond is formed.

The preceding examples illustrate also that reagents tend to be selective. Thus, in Eq. (7), hydroxide ion attacks the carbon-bromine bond rather than the carbon-carbon bond. Or in Eq. (8), iodide ion attacks the carbon-iodine bonds rather than the more numerous carbon-hydrogen bonds. The selectivity is at least in part determined by the energy required to break the old bond, as well as by the energy released when the new bond is formed. The reagent tends to attack that site in the molecule at which reaction can take place with the least expenditure of energy.

The selectivity of reagents for particular types of bonds enables us to predict the molecular structure of the product even for reactions involving very complicated molecules, since only a small portion of the original molecule is altered.

Some important types of reactions

Although myriads of reactions are known involving covalent molecules, it is possible to group them into a few broad classifications. Some of the more important classifications will be considered in this section.

THE SUBSTITUTION REACTION

In the substitution reaction, one atom or group of atoms in the molecule is replaced by another atom or group of atoms. The net result is the breaking of one covalent bond and the formation of a new one in its place. The following reactions are typical substitution reactions:

1. The reaction of ethyl bromide with hydroxide ion; this reaction was discussed on p. 288 and is shown in Eq. (7).
2. The bromination of benzene to form bromobenzene.

$$|\overline{\underline{Br}}-\overline{\underline{Br}}| + C_6H_6 \longrightarrow C_6H_5\overline{\underline{Br}}| + H-\overline{\underline{Br}}| \qquad (9)$$

Benzene, C_6H_6 Bromobenzene, C_6H_5Br

THE DECOMPOSITION REACTION

The decomposition of a covalent molecule involves its breakdown into smaller molecules or into atoms. Usually this is accomplished by heating the substance, although other methods, such as electron bombardment (p. 252), are also employed. As the temperature is raised, the fraction of molecules with energy sufficient to undergo bond rupture becomes appreciable. Even at ordinary temperatures, the atoms in a molecule are vibrating with respect to one another. As the energy of the molecule increases, the violence of these vibrations becomes sufficient to cause the weakest bond to rupture. This phenomenon is depicted in Fig. 16-2 for a diatomic molecule. An example is gaseous bromine, which breaks up into the free atoms at elevated temperatures.

$$|\overline{\underline{Br}}-\overline{\underline{Br}}| \xrightarrow{\text{Heat}} 2\,|\overline{\underline{Br}}\,\boxed{\cdot} \qquad (10)$$

Dinitrogen tetroxide (N_2O_4) on heating dissociates into two molecules of nitrogen dioxide (NO_2).

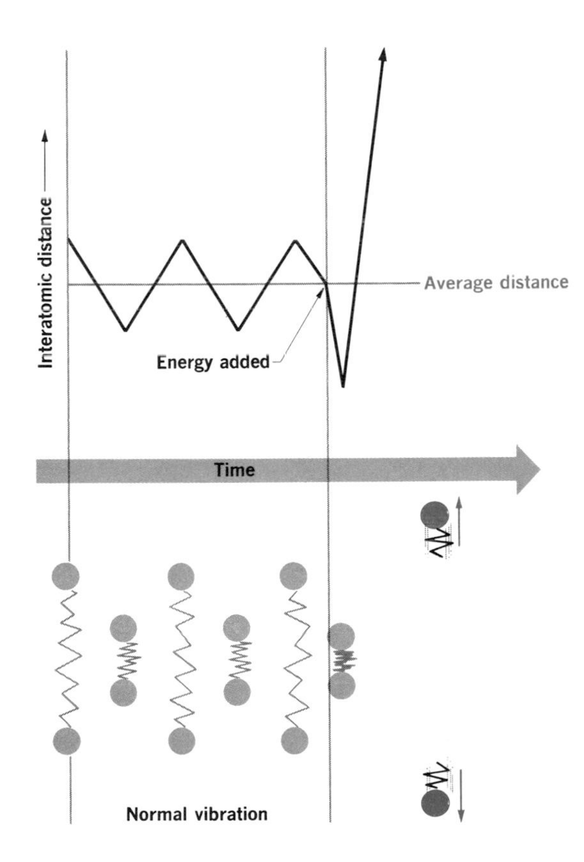

Fig. 16-2
Bond rupture due to increased amplitude of vibration caused by addition of energy.

```
|Ō|         |O|                    |O|              |O|
   \        //                    //               //
    N—N          Heat ⟶     [·]N        +   [·]N               (11)
   //   \                         \               \
|O|      |O|                      |O|              |O|
Dinitrogen                    Nitrogen dioxide
tetroxide
```

Other decomposition reactions are more complicated. For example, when phosphorus pentachloride is heated, it decomposes to give phosphorus trichloride and chlorine.

```
|Cl  |Cl|                                 |Cl|
    \  |                                    |
     P—Cl|    ⟶    |Cl—Cl|   +   |P—Cl|                      (12)
    /  |                                    |
|Cl  |Cl|                                 |Cl|
Phosphorus                             Phosphorus
pentachloride                          trichloride
```

When *iso*-propyl alcohol is passed through a hot tube containing solid aluminum oxide, the products are propylene and water.

```
     H   H   H                 H   H   H
     |   |   |                 |   |   |
 H—C—C—C—H     ⟶     H—C—C=C—H   +   H—Ō—H          (13)
     |   |   |                 |
     H  |O|  H                 H
         |
         H
Iso-propyl alcohol            Propylene
```

A commercially important decomposition reaction is the "cracking" of long-chain hydrocarbons to produce smaller molecules suitable for use as motor fuel.

THE ELIMINATION REACTION

Some of the preceding examples may be referred to as *elimination reactions* rather than decomposition. An elimination reaction generally produces a double or triple bond by the removal of atoms or groups from adjacent sites in the molecule. For example, in the decomposition of *iso*-propyl alcohol, a hydroxyl group and a hydrogen atom are detached from adjacent carbon atoms.

$$\begin{array}{c} \text{H}\quad\text{H}\quad\text{H} \\ |\quad\ |\quad\ | \\ \text{H—C—C—C—H} \\ |\quad\ |\quad\ | \\ \text{H}\ \ \boxed{\text{O}}\ \ \boxed{\text{H}} \\ | \\ \text{H} \end{array}$$

The result is that a water molecule is formed, and the C—C single bond becomes a C=C double bond.

Frequently, the elimination is assisted by the presence of a reagent which combines with the atoms or groups that are eliminated. Thus 1,2-dichloroethylene is converted to acetylene in the presence of metallic zinc.

$$\underset{\text{Trans-1,2-dichloroethylene}}{\text{ClHC=CHCl}} + \text{Zn} \longrightarrow \underset{\text{Acetylene}}{\text{H—C}\equiv\text{C—H}} + \underset{\text{Zinc chloride}}{\text{ZnCl}_2} \qquad (14)$$

Zinc combines with the chlorine atoms to form zinc chloride.

THE ADDITION REACTION

The addition reaction can be regarded as a reversal of the elimination reaction. Two atoms or groups of atoms are added to adjacent sites in a molecule containing a double or a triple bond. For example, ethylene reacts with bromine to form ethylene dibromide.

$$\text{H—CH=CH—H} + \text{Br—Br} \longrightarrow \underset{\text{Ethylene dibromide}}{\text{H—CHBr—CHBr—H}} \qquad (15)$$

Two bromine atoms add to the adjacent carbon atoms, thus converting the double bond into a single bond.

REARRANGEMENT

The conversion of one structural isomer into another is referred to as a *rearrangement reaction.* For example, *normal*-butane is converted to its structural isomer, *iso*-butane, in the presence of a catalyst, aluminum chloride.

$$\begin{array}{ccccccccc} & H & & H & & H & & H & \\ & | & & | & & | & & | & \\ H- & C & - & C & - & C & - & C & -H \\ & | & & | & & | & & | & \\ & H & & H & & H & & H & \end{array} \xrightarrow[\text{Catalyst}]{AlCl_3} \begin{array}{ccccccc} & & & H & & & \\ & & & | & & & \\ & & & H-C-H & & & \\ & H & & | & & H & \\ & | & & | & & | & \\ H- & C & - & C & - & C & -H \\ & | & & | & & | & \\ & H & & H & & H & \end{array} \qquad (16)$$

Normal-butane   *Iso*-butane

Another example, which involves an even more extensive rearrangement of the original molecule, is the conversion of *neo*-pentyl alcohol to *tertiary*-amyl alcohol in the presence of an acid.

$$\begin{array}{ccccccccc} & & & H & & & & & \\ & & & | & & & & & \\ & & & H-C-H & & & & & \\ & H & & | & & H & & & \\ & | & & | & & | & & & \\ H- & C & - & C & - & C & - & \overline{\underline{O}} & -H \\ & | & & | & & | & & & \\ & H & & | & & H & & & \\ & & & H-C-H & & & & & \\ & & & | & & & & & \\ & & & H & & & & & \end{array} \xrightarrow{\text{Acid}} \begin{array}{ccccccccc} & & & H & & & & & \\ & & & | & & & & & \\ & & & H-C-H & & & & & \\ & H & & | & & H & & H & \\ & | & & | & & | & & | & \\ H- & C & - & C & - & C & - & C & -H \\ & | & & | & & | & & | & \\ & H & & |O| & & H & & H & \\ & & & | & & & & & \\ & & & H & & & & & \end{array} \qquad (17)$$

Neo-pentyl alcohol   *Tertiary*-amyl alcohol

The net result of this rearrangement is that a CH_3 group and an OH group have exchanged places.

CHAIN REACTIONS

Chlorine molecules, when irradiated with ultraviolet light, absorb enough energy to break the chlorine-chlorine bond. This is an example of a decomposition reaction that is initiated photochemically. When chlorine is irradiated in the presence of methane, the chlorine atoms thus produced will attack the C—H bond, forming HCl and ejecting the methyl group. These reactions are shown in Eqs. (18) and (19).

DISSOCIATION OF CHLORINE:

$$|\overline{\underline{Cl}}-\overline{\underline{Cl}}| \xrightarrow[\text{irradiation}]{\text{Ultraviolet}} 2|\overline{\underline{Cl}}\,\boxed{\cdot} \tag{18}$$

SUBSTITUTION OF Cl FOR CH_3:

$$|\overline{\underline{Cl}}\,\boxed{\cdot} + H-\overset{\displaystyle H}{\underset{\displaystyle H}{\overset{|}{\underset{|}{C}}}}-H \longrightarrow |\overline{\underline{Cl}}-H + \boxed{\cdot}\,\overset{\displaystyle H}{\underset{\displaystyle H}{\overset{|}{\underset{|}{C}}}}-H \tag{19}$$

The methyl molecule (CH_3) is unstable, since it has a half-filled orbital, and will attack another molecule of chlorine, thereby producing a molecule of methyl chloride and another free chlorine atom.

$$H-\overset{\displaystyle H}{\underset{\displaystyle H}{\overset{|}{\underset{|}{C}}}}\,\boxed{\cdot} + |\overline{\underline{Cl}}-\overline{\underline{Cl}}| \longrightarrow H-\overset{\displaystyle H}{\underset{\displaystyle H}{\overset{|}{\underset{|}{C}}}}-\overline{\underline{Cl}}| + \boxed{\cdot}\,\overline{\underline{Cl}}| \tag{20}$$

The chlorine atom thus produced can attack another methane molecule, according to Eq. (19). This leads to the formation of yet another methyl radical capable of reacting according to Eq. (20).

The cycle consisting of Eqs. (19) and (20) can be repeated many times. It has been found that the decomposition of one molecule of chlorine by radiant energy will trigger as many as ten thousand cycles, that is, ten thousand molecules of methyl chloride and of hydrogen chloride are produced as a result of the initial splitting of a single chlorine molecule. A series of reactions in which some outside agency initiates a reaction which then repeats itself again and again, is called a *chain reaction.*

The foregoing examples represent a few of the reactions to be found in each of the major classifications. They involve relatively simple molecules, but the reader should realize that the same reactions will occur also with vastly more complicated molecules. Furthermore, several of these reactions may be going on simultaneously in the same reaction mixture. When *iso*-propyl bromide is treated with hydroxide ion, for example, both elimination and substitution occur. This results in a mixture of products.

$$
\text{H—}\underset{\text{H}}{\overset{\text{H}}{\text{C}}}\text{—}\underset{|\underline{\text{Br}}|}{\overset{\text{H}}{\text{C}}}\text{—}\underset{\text{H}}{\overset{\text{H}}{\text{C}}}\text{—H} \xrightarrow[\text{OH}^-]{\text{Hydroxide + ion}}
$$

Iso-propyl bromide

Elimination $\longrightarrow$ $\text{H—}\underset{\text{H}}{\overset{\text{H}}{\text{C}}}\text{—}\overset{\text{H}}{\text{C}}\text{=}\overset{\text{H}}{\text{C}}\text{—H}$ (21)

Propylene

Substitution $\longrightarrow$ $\text{H—}\underset{\text{H}}{\overset{\text{H}}{\text{C}}}\text{—}\underset{|\text{O}|\text{—H}}{\overset{\text{H}}{\text{C}}}\text{—}\underset{\text{H}}{\overset{\text{H}}{\text{C}}}\text{—H}$ (22)

Iso-propyl alcohol

The relative proportion of the products depends on the reaction conditions.

The speed of chemical reactions

Reactions of covalent molecules require anywhere from a fraction of a second to days to be completed. For any specific reaction, the time required depends on the conditions under which the reaction is carried out. The chemist tries to adjust matters so that the reaction is neither explosively fast nor inconveniently slow. He also tries to find conditions under which the desired product is isolated in good yield and in a high state of purity, that is, conditions under which the desired reaction goes virtually to completion, and little of the reagents are wasted because of the formation of undesired by-products.

Perhaps the most important controllable factor is the *reaction temperature.* Almost without exception, an increase in temperature results in acceleration of the reaction, and in many cases the acceleration is very marked. Thus, by carrying out the reaction in an ice bath instead of at room temperature, the reaction may often be slowed down to one-tenth of its original speed; and by raising the temperature to 100 °C, the rate may quite possibly increase a hundred-fold.

Another important factor is the *concentration* of the reactants. The concentration is a measure of the number of molecules in unit volume and is usually expressed in moles (gram molecular weights) per liter. In the gas phase, the concentration of molecules is greater, the greater the pressure (see Ch. 5). In the liquid phase, the concentration of a substance may be varied by dissolving more or less of it in a given amount of solvent. In both cases an increase in the concentration of the reactants frequently results in an increased rate of reaction.

Still another way of changing the rate of a reaction is to carry out the reaction in the presence of a catalyst, a substance that takes part in the reaction and thereby accelerates it without, however, changing the nature of the products.

The magnitude of these various factors is best illustrated by some numerical data. First, let us consider the reaction between methyl iodide gas (CH_3I) and hydrogen iodide gas (HI), to produce methane (CH_4) and iodine (I_2):

$$CH_3I + HI \longrightarrow CH_4 + I_2 \tag{23}$$

For this reaction, it is possible to vary the initial concentrations of the two reactants as well as the reaction temperature.

To convey an idea of the speed of this reaction as one varies the temperature and the concentration, it is convenient to record the half-reaction time, that is, the time required for one-half of the original reactants to disappear. The results of a number of experiments are shown in the table.

REACTION OF CH_3I WITH HI

Initial concentration		Reaction	Half-reaction
CH_3I	HI	temperature	time
	(a) Effect of changing the temperature		
0.1 mole/liter	0.1 mole/liter	100°C	7.3 yr
0.1 mole/liter	0.1 mole/liter	200°C	280 min
0.1 mole/liter	0.1 mole/liter	300°C	34 sec
	(b) Effect of changing the concentration		
0.01 mole/liter	0.01 mole/liter	300°C	340 sec
0.1 mole/liter	0.1 mole/liter	300°C	34 sec
1 mole/liter	1 mole/liter	300°C	3.4 sec

Examination of part (a) of the table reveals a dramatic decrease in the half-reaction time as the temperature increases. Whereas at 100 °C the reaction is much too slow to be of practical value, requiring years to go to completion, at 300 °C the reaction proceeds at quite a convenient rate. The reaction is also speeded up by the increase of the concentrations of the reactants, as shown in part (b) of the table, although the effect is not nearly so marked.

The preceding reaction involves the interaction of two different kinds of molecules, CH_3I and HI. It is also of interest to examine a reaction that involves the breakdown of a single molecular species, such as a decomposition reaction. An example is furnished by the decomposition of gaseous *iso*-propyl bromide (C_3H_7Br) into propylene gas (C_3H_6) and hydrogen bromide gas (HBr).

$$\underset{\textit{Iso-propyl bromide}}{H-\underset{H}{\overset{H}{C}}-\underset{|\underline{Br}|}{\overset{H}{C}}-\underset{H}{\overset{H}{C}}-H} \longrightarrow \underset{\text{Propylene}}{H-\underset{H}{\overset{H}{C}}-\overset{H}{C}=\overset{H}{C}-H} + \underset{\text{Hydrogen bromide}}{H-\overline{Br}|} \qquad (24)$$

Some experimental data obtained for this reaction are shown in the following table.

DECOMPOSITION OF GASEOUS *ISO*-PROPYL BROMIDE

Initial concentration of C_3H_7Br	Reaction temperature	Half-reaction time
	(a) Effect of changing the temperature	
0.1 mole/liter	200°C	6.6 yr
0.1 mole/liter	300°C	8.2 hr
	(b) Effect of changing the concentration	
0.001 mole/liter	300°C	8.2 hr
0.01 mole/liter	300°C	8.2 hr
0.1 mole/liter	300°C	8.2 hr

Again there is a marked increase in the speed of the reaction as the temperature is raised, but the half-reaction time is now independent of the initial concentration, since only one kind of molecule is involved.

The speeding up of reactions by catalysts is of great practical importance. For example, a possible process for the manufacture of sulfuric acid involves the reaction of sulfur dioxide (SO_2) with oxygen (O_2) to form sulfur trioxide (SO_3). The latter subsequently reacts with water to form sulfuric acid.

$$O_2 + 2SO_2 \longrightarrow 2SO_3 \tag{25}$$

$$SO_3 + H_2O \longrightarrow H_2SO_4 \tag{26}$$

Ordinarily, reaction (25) is inconveniently slow. However, the reaction is speeded up by the addition of nitric oxide (NO). A possible interpretation is that nitric oxide reacts rapidly with oxygen to form nitrogen dioxide (NO_2), which in turn reacts rapidly with sulfur dioxide to form the desired sulfur trioxide.

$$O_2 + 2NO \longrightarrow 2NO_2 \tag{27}$$

$$NO_2 + SO_2 \longrightarrow SO_3 + NO \tag{28}$$

Furthermore, as a result of reaction (28), the nitric oxide is regenerated and can repeat its catalytic action. An over-all increase in the rate of the reaction is thus achieved by substituting two fast reactions (27 and 28) for one slow one (25).

Another device for speeding up a reaction is to add a *surface catalyst.* This is a solid on the surface of which the reactants are adsorbed in such a way that they are more ready to react than they would be in the absence of the solid. For example, the decomposition of *iso*-propyl alcohol to propylene and water [Eq. (13)] takes place much more readily in the presence of solid aluminum oxide than in its absence.

$$\begin{array}{c} \mathrm{H\ \ H\ \ H} \\ \mathrm{H{-}C{-}C{-}C{-}H} \\ \mathrm{H\ |\overline{O}|\ H} \\ \mathrm{H} \end{array} \xrightarrow[\text{oxide catalyst}]{\text{Aluminum}} \begin{array}{c} \mathrm{H\ \ H\ \ H} \\ \mathrm{H{-}C{-}C{=}C{-}H} \\ \mathrm{H} \end{array} + \mathrm{H{-}\underline{\overline{O}}{-}H} \tag{13}$$

Surface catalysis is a highly specific phenomenon. Starting with a given set of reactants, different catalysts can accelerate different reactions leading to different products. Thus, if copper is used instead of aluminum oxide, *iso*-propyl alcohol does not decompose to propylene and water but decomposes in a completely different manner to form acetone (C_3H_6O) and hydrogen.

$$\begin{array}{ccccccc} & H & & H & & H & \\ & | & & | & & | & \\ H- & C & - & C & - & C & -H \\ & | & & | & & | & \\ & H & & |\underline{O}| & & H & \\ & & & | & & & \\ & & & H & & & \end{array} \xrightarrow[\text{220–270°C}]{\text{Copper catalyst}} \underset{\text{Acetone}}{\begin{array}{ccccccc} & H & & |\overline{O}| & & H & \\ & | & & \| & & | & \\ H- & C & - & C & - & C & -H \\ & | & & & & | & \\ & H & & & & H & \end{array}} + \; H-H \qquad (29)$$

This example illustrates one of the important problems of practical chemistry. The given set of reactants may quite possibly be capable of reacting in different ways to form alternative sets of products. The chemist must find conditions under which the desired products are formed rapidly and the undesired ones are formed relatively slowly.

SPEED OF REACTIONS AND CHEMICAL EQUILIBRIUM

It is clear from the preceding examples that chemical equilibrium and the speed of chemical reactions are unrelated phenomena. The addition of a catalyst can increase the speed dramatically, but it cannot affect the composition of the final equilibrium mixture. On the other hand, much chemistry is done on mixtures that are not at equilibrium. When several reactions are possible but their speeds are controlled so that only one of the possible products is formed rapidly, the mixture from which that product is isolated is obviously not at equilibrium. However, the mixture may *appear* to be at equilibrium if the speeds at which the other products are formed are very slow.

A mixture of substances, actually not at equilibrium, is said to be in *metastable equilibrium* if its progress towards a state of genuine equilibrium is imperceptibly slow. Chemistry owes much of its fascinating diversity to the fact that metastable equilibrium is possible. Most of the articles of daily life—clothing, books, furnishings, anything that is combustible—would not last long enough to be useful were it not for metastable equilibrium. Indeed, life itself would be impossible, because all organic matter is metastable in the presence of oxygen-containing air.

Theory of reaction rates

We wish to describe briefly a theory that explains the effect of temperature, concentration, and catalysts on the rates of reactions in which two or more molecules must come together. According to the theory, the rate of the reaction, that is, the number of molecules reacting in unit time, is equal to the product of the following independent factors:

1. The total number of collisions in unit time.
2. The probability that any given collision will bring the reaction zones of the two reagent molecules into direct contact.
3. The probability that the colliding molecules have sufficient energy to form an activated complex.

We shall now consider how these factors vary with the reaction conditions.

THE TOTAL NUMBER OF COLLISIONS IN UNIT TIME

The number of collisions experienced in unit time by a molecule in the gas phase was discussed in Ch. 5 (p. 81). It was seen that this number is very large, and that at a given temperature it increases with the pressure, which in turn is a measure of the concentration of the gas molecules. The same rule applies to a mixture of two or more gases. The number of collisions between any two kinds of molecules increases when the concentration of either of the two substances increases. One would therefore expect that the rate of a reaction between the two increases also. The reaction between methyl iodide and hydrogen iodide, discussed on p. 297, furnishes a typical example. An increase in the concentration of the reactants results in an increased reaction rate and a decreased half-reaction time. Many reactions occurring in liquid solution behave in much the same way.

The number of collisions varies not only with the concentrations of the colliding molecules, but also with the temperature. As the temperature is raised and the average velocity of the molecules increases, collisions become more frequent even while the concentrations remain constant. Hence, we would expect an increase in the rate of the reaction. This is again in agreement with observation.

COINCIDENCE OF REACTION ZONES

Reaction will occur only in those collisions in which the reaction zones of the two molecules come in direct contact. Since the reaction zone is normally only a small portion of the total molecule, the odds are that only a fraction of all collisions will meet this requirement.

ENERGY REQUIREMENT

Just as the boundary of an atom is defined by its outermost electron orbitals, so the boundary of a molecule is defined by the outermost electron orbitals of the molecule. Owing to the rapid motion of the electrons in these orbitals, we may imagine that all atoms and molecules are surrounded by a region of negative charge. This region will act as a defensive barrier against attack by another molecule, since the negative charge will repel the negative charge on the surface of the approaching molecule. In a normal collision this barrier is not penetrated; the molecules come together, collide, and move apart.

In order that reaction may occur, the attacking reagent must penetrate this defensive barrier and come within bond-forming distance of the molecule with which it is to react. This will happen only if the colliding molecules are moving fast enough to overcome the mutual repulsion of their outer boundaries. Moreover, in order for reaction to occur, there must be enough energy so that the relatively unstable configuration of the activation complex can be attained. For the great majority of reactions, the activation energy thus required is considerably greater than the kinetic energy possessed by the average molecule.

It has been pointed out (p. 79) that only a small fraction of all the molecules will at any given instant possess energy well in excess of the average and, therefore, be capable of reacting. Of the collisions involving these energetic molecules, only those having the proper alignment of reaction zones will lead to the formation of product molecules. Thus, we must conclude that only a minute fraction of all collisions are effective.

Since the number of highly energetic molecules increases quite sharply with increasing temperature, as shown in Fig. 5-7, we would expect that the number of collisions with the required energy would also increase quite sharply. As a result, there should be a sharp increase in the rate of the reaction with increasing temperature. This is in agreement with experiment, as shown in the tables on pp. 297–298.

Suggestions for further reading

Campbell, J. A., *Why Do Chemical Reactions Occur?* (Englewood Cliffs, N. J.: Prentice-Hall, Inc., 1965).

Clark, B. F. C., and Marcker, K. A., "How Proteins Start," *Scientific American,* January 1968, p. 36.

Faller, L., "Relaxation Methods in Chemistry," *Scientific American, 220,* May 1969, p. 30.

King, E. L., *How Chemical Reactions Occur* (New York: W. A. Benjamin Co., Inc., 1963).

Kornberg, A., "The Synthesis of DNA," *Scientific American,* October 1968, p. 64.

Natta, G., "How Giant Molecules Are Made," *Scientific American,* Sept. 1957, p. 98.

Roberts, J. D., "Organic Chemical Reactions," *Scientific American,* Nov. 1957, p. 117.

Questions

1. It is stated in the text that "a reaction mechanism shares with any good theory the characteristic that it can be tested by experiment." How would you test whether:
(*a*) Acetic acid is a catalyst for the decomposition of hydrogen peroxide (H_2O_2) in aqueous solution? The chemical equation for that reaction is, $2H_2O_2 \longrightarrow 2H_2O + O_2$.
(*b*) The reaction of chlorine with methane (Eqs. 18-20) is initiated photochemically?
(*c*) The reaction of ethyl bromide with hydroxide ion (Eq. 7) requires the direct collision of these two molecules?

2. In the photochemical reaction of chlorine with methane, if the presence of chlorine atoms and methyl (CH_3) molecules in the reaction mixture could be demonstrated by direct observation, would the status of the reaction mechanisms shown in Eq. 18-20 change from microscopic theory to established fact? Explain carefully.

3. In what sense do all reactions require the input or absorption of energy? What is this energy called?

4. Explain the meaning, and discuss the practical implications, of the concept of *least structural change.*

5. Classify each of the following reactions as substitution, decomposition, elimination, addition, or rearrangement.

(*a*) $$CH_3C{\equiv}CH + H_2O \longrightarrow CH_3CH_2\overset{\displaystyle H \atop |}{C}{=}O$$

(*b*) $$CH_3CH_2OH \xrightarrow[\text{catalyst}]{\text{heat}} CH_2{=}CH_2 + H_2O$$

(*c*) $$\begin{array}{ccc} & CH_2 & \\ / & & \backslash \\ HC & & CH_2 \\ \| & & | \\ HC & & CH_2 \\ \backslash & & / \\ & CH_2 & \end{array} \xrightarrow[\text{selenium catalyst}]{\text{heat}} \begin{array}{ccc} & CH & \\ / & & \backslash\backslash \\ HC & & CH \\ \| & & | \\ HC & & CH \\ \backslash & & // \\ & CH & \end{array} + 2H_2$$

(*d*) $$CH_3C{\equiv}C\underset{\displaystyle | \atop H}{\overset{\displaystyle H \atop |}{C}}{-}OH \longrightarrow CH_3\underset{\displaystyle \| \atop O}{C}{-}\overset{\displaystyle H \atop |}{C}{=}\underset{\displaystyle | \atop H}{\overset{\displaystyle H \atop |}{C}}$$

6. Write equations, employing structural formulas, for each of the following reactions.

(*a*) $$\begin{array}{ccc} Cl & & Cl \\ & C{=}C & \\ H & & H \end{array} + Zn,$$ in an elimination reaction.

(*b*) $$H{-}\underset{\displaystyle | \atop H}{\overset{\displaystyle H \atop |}{C}}{-}\underset{\displaystyle | \atop Br}{\overset{\displaystyle H \atop |}{C}}{-}\overset{\displaystyle H \atop |}{C}{=}\underset{\displaystyle | \atop Br}{\overset{\displaystyle H \atop |}{C}} + Br_2,$$ in an addition reaction.

(*c*) $$H{-}\underset{\displaystyle | \atop H}{\overset{\displaystyle H \atop |}{C}}{-}\underset{\displaystyle | \atop H}{\overset{\displaystyle H \atop |}{C}}{-}OH + Na,$$ in a reaction that

results in the formation of H_2. Should the reaction be classified as substitution or as oxidation-reduction?

(*d*) $$H{-}\underset{\displaystyle | \atop H}{\overset{\displaystyle H \atop |}{C}}{-}OH + HI,$$ in a reaction similar to that shown for ethyl alcohol on p. 247. Should the reaction be classified as substitution? Is a knowledge of the reaction mechanism required before a classification can be made?

7. From an examination of the data for the decomposition of *iso*-propyl bromide (p. 298), if one-half of the reaction at 300°C is consumed in 8.2 hrs., how long will it take for three-quarters (75%) to be consumed? For *all* of it to be consumed?

8. From an examination of the data for the reaction of CH_3I with HI (p. 297), which would you say is more important in determining the speed of the reaction, the concentration or the temperature? Can you offer an explanation?

9. The reaction, $H_2 + D_2 \longrightarrow 2HD$, in the gas phase at moderate temperatures probably occurs by the direct collision of a molecule of H_2 with a molecule of D_2. What will be the effect on the rate of this reaction if:
(*a*) The concentration of both reactants is increased.
(*b*) The temperature is reduced.
(*c*) The total pressure is increased.
(*d*) A catalyst is added.
(*e*) The reaction mixture is stirred.

10.* The reaction, $H_2 + D_2 \longrightarrow 2HD$, is often discussed in terms of a reaction mechanism of the type:

$$\begin{matrix} H & & D \\ | & + & | \\ H & & D \end{matrix} \longrightarrow \begin{matrix} H \cdots D \\ \vdots \quad \vdots \\ H \cdots D \end{matrix} \longrightarrow \begin{matrix} H—D \\ + \\ H—D \end{matrix}$$

activated complex

Using electron-dot formulas, try to represent the electronic structure of the activated complex.

11. What are the macroscopic characteristics of a chain reaction?

12. With reference to Fig. 5-7, consider the reaction, $N_2 + 3H_2 \longrightarrow 2NH_3$. Assuming that a nitrogen molecule needs a speed of at least 4500 feet per second to undergo reaction with hydrogen, what fraction of the N_2 molecules will be able to react at 0°C? At 1000°C?

13. Why would you expect complex molecules containing many atoms to react more slowly than simple molecules, all other things being equal?

* Questions marked with an asterisk (*) are more difficult or require supplementary reading.

If we begin in certainties,
we shall end in doubts;
but if we begin with doubts,
and are patient in them,
we shall end in certainties.

FRANCIS BACON (1561–1626)

Seventeen

THE ATOMIC NUCLEUS

THE EXPERIMENTS of Ernest Rutherford on the scattering of alpha particles by thin foils of metal offered strong evidence for the existence of an atomic nucleus bearing a positive charge of electricity and containing almost the entire mass of the atom. In our discussion of chemical bonding and molecular structure, the nucleus played only a minor role, because these phenomena could be discussed fruitfully by considering only the electrons in the outermost shell of the atom. There is, however, a whole range of extremely important phenomena whose understanding depends on a detailed knowledge of the nucleus. In this chapter, we shall consider the constitution of the atomic nucleus and examine some of the ideas that have been developed concerning its structure.

The discovery of the proton

Recall that radioactivity involves the *spontaneous* disintegration of atomic nuclei with the ejection of small particles, such as alpha particles (helium nuclei) or beta particles (electrons). Rutherford, while studying natural radioactivity, became convinced that the disintegration was due to the intrinsic instability of the radioactive nucleus. He then began to speculate on the possibility of bringing about nuclear disintegration by artificial means. In his early experiments, he bombarded the nuclei of certain elements with energetic alpha particles ejected by radioactive atoms in the course of their spontaneous disintegration, thus hoping to cause new nuclear reactions to take place.

The experiment that interests us at this point is the bombardment of nitrogen nuclei with alpha particles. Rutherford found, in 1918, that when nitrogen gas is exposed to a stream of fast-moving alpha particles, every once in a while (about once for every 10,000 alpha particles) the nucleus of a nitrogen atom disappears, and in its place there appears an atom of oxygen and another less-heavy particle with a charge of $+1$ and a mass of 1. The reader will recognize that the latter particle is a hydrogen nucleus or *proton.* (This experiment was carried out in a Wilson cloud chamber, the operation of which will be considered shortly.)

This nuclear reaction is of tremendous interest for two reasons. First, and perhaps most startling, is the fact that it represents the artificial *transmutation* of one element into another! Never before in all his tinkering with matter had man been able to transform one element into another. During the Middle Ages the alchemists had as their primary goal the conversion of the base metals into gold, but of course they never succeeded. In all *ordinary chemical reactions* the nucleus of the atom is left unaltered, and therefore, the element survives as such throughout the most extensive and complicated of chemical reactions; but in Rutherford's experiment, the alpha particle collided directly with the nucleus of the nitrogen atom, and a new nucleus, that of the oxygen atom, was produced.

The second and more fundamental reason for our interest in this experiment is that the observation of the proton gives us a clue as to the possible composition of the nucleus. It was later found that this same particle is produced in many other nuclear reactions. It is therefore probable that the proton is one of the fundamental building blocks of the nucleus, in the same way that the electron is one of the fundamental building blocks of the entire atom.

The discovery of the neutron

A number of years after Rutherford's discovery of the proton, Chadwick, a former student of Rutherford's, demonstrated the existence of yet another fundamental particle.† In 1930 it was observed that, when beryllium was bombarded with alpha particles, there was produced a very penetrating radiation which, for a time, was thought to be gamma radiation. Chadwick demonstrated, however, that this mysterious radiation was a stream of fast-moving particles of about the same mass as a proton but having no electric charge. Because of their electrical neutrality, these particles were called *neutrons*.

The lack of charge on the neutron is responsible for its great penetrating power. Protons of the same energy are stopped by a much thinner layer of matter because of the interaction of their electric charges with the charges of the nuclei and electrons in the matter they are traversing.

In discussing nuclear particles, it is convenient to introduce the concept of the *mass number* (A). The mass number is defined as the atomic mass of the nuclear particle or nucleus (in atomic weight units), rounded off to the nearest whole number. Thus, the mass number of the neutron, as well as that of the proton, is *one,* whereas that of the alpha particle is *four.*

The proton-neutron theory of the nucleus

With the discovery of the proton and the neutron, it became possible to develop a consistent theory of nuclear composition, according to which nuclei are made up of protons and neutrons. Since each proton has a single positive charge, the number of protons in the nucleus of a given element is equal to the total nuclear charge characteristic of that element, which in turn is equal to the atomic number (p. 136). For example, the element of atomic number 10 has a nuclear charge of +10 and, therefore, contains 10 protons in the nucleus.

However, with the exception of the hydrogen nucleus, the ob-

† A "fundamental particle" is a particle that cannot be further broken down into still simpler particles. This definition is reminiscent of our earlier definition of the elements; and in a certain sense, of course, the concept of a small number of fundamental particles replaces the concept of a small number of different kinds of atoms. In this chapter, four particles are mentioned: electrons, protons, neutrons, and mesons. However, within the past few years a large number of other particles has been discovered; so many, in fact, that the "fundamental" identity of most of them is now under question.

served masses of all known nuclei are greater than the total mass of the required number of protons. The theory must therefore make provision for adding mass to the nucleus without adding charge. This is done by adding a number of neutrons of unit mass and zero charge. For example, if an atom of atomic number 10 has a mass number of 20, ten neutrons must be added to the 10 protons. In general, the number of neutrons present in the nucleus of any atom is equal to the *difference* between the mass number (A) and the atomic number (Z).

The proton-neutron theory of the nucleus is summarized by the following relationships:

Number of protons = Atomic number (Z) (1)
Number of neutrons = Mass number (A) − Atomic number (Z) (2)

The reader will recall that the number of orbital electrons in the *neutral atom* is equal to the nuclear charge. Therefore, we obtain an additional relationship:

Number of orbital electrons = Number of protons in nucleus = Z (3)

The application of the proton-neutron theory to a number of different nuclei is illustrated in the table.

APPLICATION OF PROTON-NEUTRON THEORY

Nucleus	Mass number (A)	Atomic number (Z)	No. of protons	No. of neutrons
Hydrogen-1	1	1	1	0
Helium-4	4	2	2	2
Lithium-7	7	3	3	4
Oxygen-16	16	8	8	8
Fluorine-19	19	9	9	10
Gold-197	197	79	79	118
Radon-222	222	86	86	136

For the lighter elements, the number of protons and neutrons is about equal, whereas for the heavy elements, the number of neutrons is significantly greater than the number of protons. Since the protons all possess a positive charge, they mutually repel one another. As the number of protons in the nucleus rises, more and more neutrons seem to be required to separate the protons and to reduce their repulsive interaction.

Isotopes

All atoms of the same element have identical nuclear charges, but they need not have identical masses. For ordinary chemical reactions this latter fact rarely ever causes any complications, because the chemical behavior of all the atoms of a given element is very similar, regardless of their masses. Thus, for chemical purposes, it is sufficient to characterize the atoms of the given element by their average atomic weight, which is the mean weight of all the atoms. For example, naturally occurring tin (atomic number 50) is a mixture of atoms with the atomic mass numbers 112, 114, 115, 116, 117, 118, 119, 120, 122, and 124. The mean weight of all the tin atoms in such a sample has been determined by chemical analysis as 118.7. This mean atomic weight can be used in all chemical calculations involving samples of tin derived from natural sources.

In contrast, in a nuclear reaction, the mass numbers of the individual nuclei are important factors. For example, under a given set of conditions, uranium atoms of mass 238 and 235 undergo entirely different nuclear reactions. It is, therefore, important to specify not only the element, but also the mass number of the *isotope* involved in the reaction.†

The two isotopes of uranium mentioned above can be represented as follows:

Uranium-235 $= {}_{92}U^{235}$

Uranium-238 $= {}_{92}U^{238}$

In this representation, the nuclear charge (Z) is shown as a subscript preceding the symbol of the element, and the nuclear mass (A) is shown

† Derived from the Greek words *iso* and *topos*, meaning "same place," and refers to atoms belonging to the same space in the periodic table. The atoms differ only in their atomic mass.

as a superscript following the symbol. Thus, for the general case we have, $_{Z}\text{Symbol}^{A}$.

Particles encountered in nuclear reactions may also be symbolized in this way. Some examples are shown in the following table.

SOME IMPORTANT NUCLEAR PARTICLES

Common name	Symbol	Mass number (A)	Charge (Z)	Remarks
Alpha particle	$_{2}He^{4}$	4	2	Helium nucleus
Beta particle	$_{-1}e^{0}$	0	−1	Electron
Proton	$_{1}H^{1}$	1	1	Hydrogen nucleus
Deuteron	$_{1}H^{2}$	2	1	Heavy hydrogen nucleus
Neutron	$_{0}n^{1}$	1	0	———

The existence of isotopes is readily understood in terms of the proton-neutron theory of the nucleus. The nuclei of different isotopes of the same element all have the same number of protons but different numbers of neutrons. Thus, $_{92}U^{235}$ has 92 protons and 143 neutrons, while $_{92}U^{238}$ has 92 protons and 146 neutrons.

By means which will be discussed in the next chapter, it has been possible to prepare a large number of artificial isotopes, so that today the 105 chemical elements exhibit over 1000 different isotopic forms. Many of these new isotopes have found widespread use as "tracers" in medicine, industry, agriculture, and in other branches of science.

The detection of nuclear particles

To study the nature of nuclear reactions such as those which led to the discovery of the proton and the neutron, it was necessary to develop techniques for the detection and characterization of nuclear particles. These techniques are different from those employed to identify the products of *chemical* reactions. In contrast to chemical reactions, where one obtains huge numbers of product molecules that can be separated by macroscopic procedures, in nuclear reactions one often attempts to detect individual particles or events. When one considers that *whole* atoms are much too small to be visible, the detection of *nuclear* particles, whose dimensions are on the order of 1/10,000 of those of atoms, must seem like an insurmountable task. Nevertheless, a variety of ingenious methods have been devised. One of these methods, the use of a scintillating screen, was discussed in connection with Rutherford's alpha scattering experiments (p. 131). There are other methods for detecting nuclear particles, some of which will even render visible the actual nuclear reactions that take place.

PHOTOGRAPHIC FILM

The initial (and accidental) discovery of radioactivity by Becquerel in 1895 depended upon the effect such particles as protons, electrons, alpha particles, and gamma rays have on a piece of photographic film. In connection with some other experiments, the French scientist had left a crystal of a uranium salt in contact with a photographic plate that was wrapped in thick black paper. Through one of those coincidences that play an important role in scientific discoveries, Becquerel developed this plate and found, much to his amazement, an image of the crystal imprinted on the plate. Evidently the uranium was emitting some invisible and penetrating radiation or particle that was capable of affecting the photographic plate in the same manner as ordinary light, but which could pass through the paper in which the plate was wrapped. This photographic technique has been refined and extended, and is widely used today in many investigations in nuclear physics. It has the obvious advantage of producing a permanent record of the passage of the nuclear particle which can be studied at leisure. Much of our knowledge, for example, of cosmic rays has come from the study of photographic plates that have been carried to high altitudes by balloons where they are exposed to the bombardment of these mysterious particles or radiations from outer space. A more mundane application is found in the so-called "film badge," which is simply a piece of film in an opaque case worn by scientists who work in the vicinity of sources of high energy radiation. Development of the film reveals the amount of radiation received by the wearer, by measurement of the amount of blackening that the film has undergone.

THE WILSON CLOUD CHAMBER

Earlier in this chapter, the transmutation of a nitrogen atom into an oxygen atom with the accompanying production of a proton was discussed. The ingenious method whereby these events, involving the nuclei of *single atoms,* were made visible and studied deserves considera-

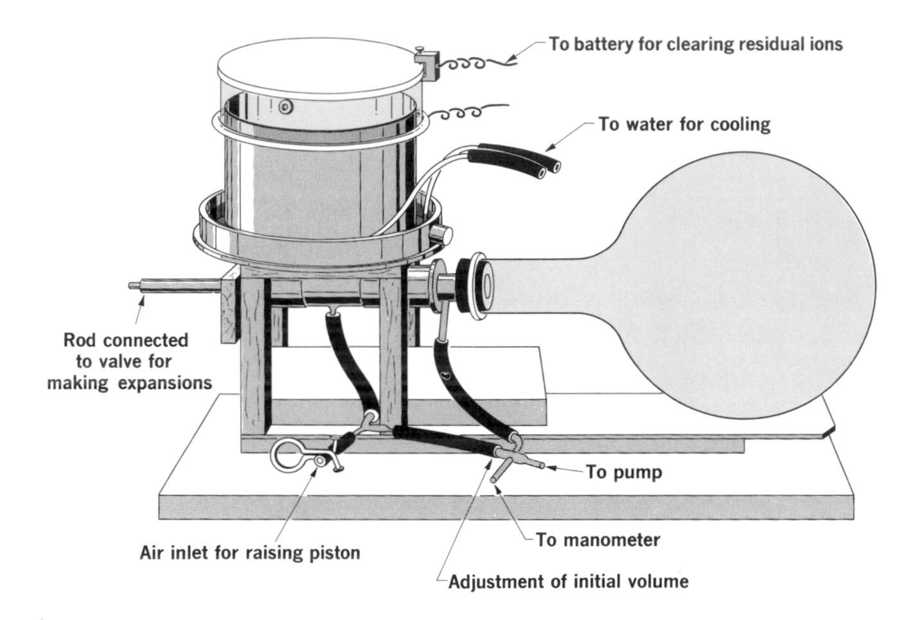

Fig. 17-1
An early model (1911) of the Wilson cloud chamber. The tracks are generated in the cylindrical chamber at the top of the apparatus. [After a photograph by P.M. S. Blackett, from *Les Inventeurs Celebres* (Paris: Mazenod, 1950).]

tion since it has been one of the most fruitful techniques for the study of nuclear properties and nuclear reactions. The device employed is called the "Wilson cloud chamber" after its inventor, C. T. R. Wilson. (See Fig. 17-1.) The manner in which the cloud chamber depicts the path of nuclear particles is in many ways analogous to the way in which vapor trails indicate the path of high-flying aircraft. Frequently, these aircraft are flying at such great altitudes as to be themselves invisible, but their passage is made quite evident by the "trail" they leave behind them. In a similar fashion, while the particle itself is not made visible in the cloud chamber, its passage is clearly marked by a "fog trail" which it leaves in its wake.

The air within the cloud chamber contains an excess of water vapor, but condensation to form a cloud does not occur under ordinary conditions because the water molecules must have some kind of centers (called condensation nuclei) around which to collect into droplets visible to the eye. Now, if a fast-moving particle like a proton or nitrogen nucleus passes through this water-laden air, it produces a wake of *ions.* These ions are formed from the nitrogen and oxygen molecules of which the air is composed by the "knocking out" of electrons by the swiftly moving nuclear particle. The ions are ideal condensation nuclei, and the water condenses on them rapidly to form visible droplets, thus producing a fog trail behind the moving nucleus. In Figs. 17-2 and 17-3 are seen fog trails (or fog tracks) produced by the passage of protons and electrons. It will be noted that the thickness and length of the tracks differ for the different particles.† The track formed by the proton is thicker because the proton produces more ions per unit length of track. These characteristic differences permit identification of the particular particle producing the fog track.

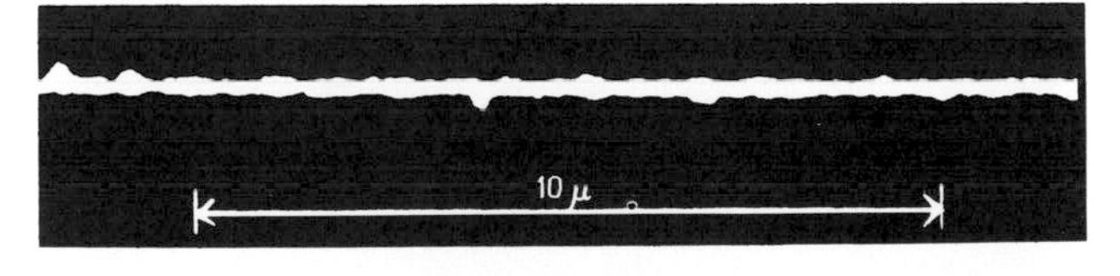

Fig. 17-2
Cloud chamber track of a slow proton.

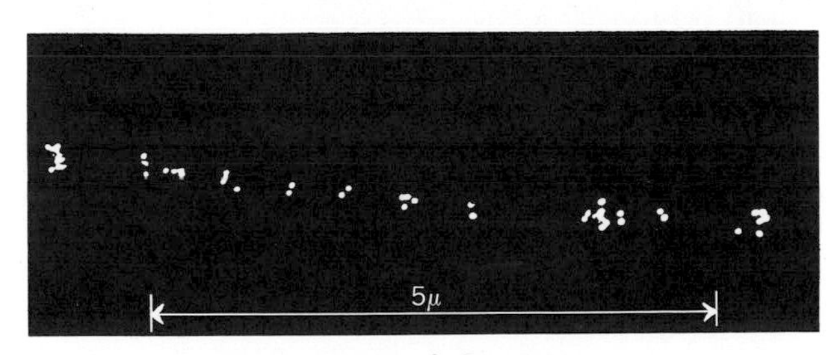

Fig. 17-3
Cloud chamber track of a fast-moving electron.

† The Greek letter μ stands for *micron,* a unit of length equal to 0.001 mm.

Fig. 17-4
A cloud chamber picture of alpha tracks in nitrogen. [Courtesy of Max Born's *Atomic Physics* (London: Blackie & Son Limited, 1946)].

Figure 17-4 is a photograph of the event described by Rutherford. Starting from the bottom of the photograph, we see the tracks left by a number of alpha particles. At the point indicated by the arrow, one of them has collided with the nucleus of a nitrogen atom; and emanating from this point are two quite different tracks, a short, thick one and a longer, less dense one. By the study of many events of this kind and the use of auxiliary apparatus, it can be shown that the short, thick track is indeed one left by an oxygen nucleus, and the longer one is due to a fast-moving proton.

THE BUBBLE CHAMBER

Because of the low density of the gas in the cloud chamber, some high-energy particles will pass through it without producing a sufficiently large number of ions to give a visible track. To overcome this difficulty, it is necessary to use a liquid that is denser and in which the range of particles is, therefore, much shorter. This modification of the cloud chamber is called a bubble chamber, and the liquid usually employed is liquid hydrogen. Upon a slight reduction of the pressure on the liquid hydrogen, bubbles form at the site of the ions and produce a track that can be photographed.

GEIGER TUBES

Geiger tubes (Fig. 17-5) also depend for their operation on the ionizing property of fast-moving, charged nuclear particles. The tube consists of a thin glass envelope through which the nuclear particles can pass.

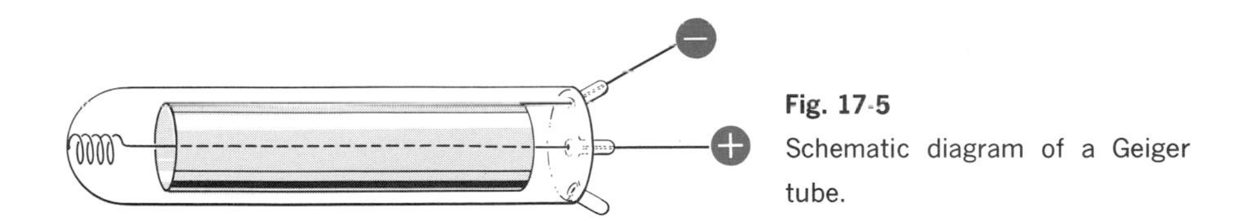

Fig. 17-5
Schematic diagram of a Geiger tube.

Inside the envelope is a cylinder of metal, down the center of which is suspended a thin wire, which is insulated from the metal cylinder. The tube is filled with a suitable gas (often a mixture of argon and ethyl alcohol vapor), and the wire and metal cylinder are connected to a source of high voltage. Ordinarily, no current flows because of the gap in the circuit owing to the lack of contact between the wire and the metal cylinder. But if a nuclear particle creates a shower of ions in this gap, a brief pulse of electricity is generated as the ions are attracted to the electrodes. This pulse can be amplified and made to activate a device that counts the pulse and records the passage of *each* particle through the detecting tube.

Nuclear structure

We have already sketched the simplest possible picture of nuclear structure: a tiny cluster of neutrons and protons at the center of the atom. This picture is, of course, quite incomplete and raises a great many new questions. For example, what holds the particles together? Are they organized in any particular way, or are they simply lumped in a random fashion? And are these the only particles to be found within the nucleus? These problems will now be considered briefly.

NUCLEAR FORCES

When we examine the characteristics of the force that binds together two protons, or a proton and a neutron, or two neutrons, or the force that binds together the 238 particles that constitute the nucleus of a uranium-238 atom, we quickly discover that this force is quite unlike anything we have encountered previously. Whereas the electrostatic forces, which were invoked to explain the formation of molecules or the binding of electrons to the positive nucleus, diminish in intensity relatively slowly as the two charged particles are separated, nuclear forces act only over *very* short distances. In fact, if the nuclear particles are separated by distances much greater than the nuclear diameter, the attraction ceases. Furthermore, this force is independent of the charge of the particles; the same force that binds together a neutron and a proton will also bind together two neutrons, or two protons.

The nature of this force is not clear. When the concept of the covalent bond was developed, an attempt was made to extend the same general idea to the field of nuclear binding. This hypothesis implied that, just as two atoms were held together by the sharing of electrons, so could two nuclear particles be held together by the exchange of electrical charge. Unfortunately, as soon as it became possible to make a somewhat accurate estimate of the strength of nuclear forces, this hypothesis became untenable. The force of attraction expected from the exchange of a positive charge between a proton and a neutron is much too weak to be responsible for binding these two particles together.

In 1935, the Japanese physicist, H. Yukawa, suggested that a then undiscovered particle, now called the *meson,* was responsible for binding the nuclear particles together. The interaction that leads to a binding force is represented in this theory as the emission by one nuclear particle and absorption by another of a particle whose mass is about one-tenth the proton mass. In support of Yukawa's hypothesis, a family of particles with masses in this range have since been discovered. Of these, particularly the pi-mesons have properties very similar to those of the particles envisioned in the theory. Therefore, this particular theory has been pursued vigorously, but it is still incapable of explaining many of the experimental facts.

It is clear that our understanding of the nature of nuclear binding forces is far from complete. Much research is being done in this area, and we may hope for exciting advances in the future.

Hideki Yukawa: 1907–
Well known for his investigations of nuclear binding forces. [Photograph courtesy Brown Brothers.]

NUCLEAR MODELS

A number of different models have been developed for nuclear structure, just as models have been developed for atomic structure and molecular structure. These nuclear models are, however, not nearly so complete as the atomic and molecular models, which describe the characteristics of atoms and molecules with considerable precision.

We shall consider first the "shell" model of the nucleus. According to this model, the nuclear particles are organized in shells, just as the electrons surrounding the nucleus are organized in shells of orbitals. The nucleus may therefore be said to have a definite structure.

The evidence in favor of a shell model for the nucleus is much like that which has led to a shell model of electronic structure. In recent years it has become apparent that many nuclear properties vary periodically in a manner similar to the periodic variation of the elements. Just as the atomic numbers 2, 10, 18, 36, and so on, are associated with elements of unusual stability, so there seem to be certain "magic numbers" of protons or neutrons which produce nuclei that are particularly stable. Thus, if the number of protons, or the number of neutrons, is 2, 8, 20, 50, and so forth, the resulting nucleus will have a relatively low

Maria Goeppert-Mayer: 1906–
A pioneer in the development of nuclear structure models. [Photograph courtesy Wide World Photos, Inc.]

internal energy, and the corresponding isotope will be relatively plentiful in nature. It is found that $_2He^4$ (2 neutrons, 2 protons) and $_8O^{16}$ (8 protons, 8 neutrons) are both much more stable than, for example, $_2He^3$ or $_8O^{17}$. When we look at the natural abundances of the isotopes, we find that definite peaks in abundance occur at the isotopes listed in the following table. Note that each of these isotopes has at least one "magic number" associated with it.

The "magic numbers" have been interpreted as representing *filled*

SOME ISOTOPES OF HIGH ABUNDANCE

Isotope	Number of protons	Number of neutrons
$_8O^{16}$	**8**	**8**
$_{20}Ca^{40}$	**20**	**20**
$_{38}Sr^{88}$	38	**50**
$_{39}Y^{89}$	39	**50**
$_{40}Zr^{90}$	40	**50**
$_{50}Sn^{118}$	**50**	68

shells of neutrons or protons, just as the numbers 2 and 10 (2 + 8) represent filled shells of electrons in atomic structure. The shell model of the nucleus has been quite fruitful in predicting the properties of new nuclei. It has been especially useful in predicting which of the nuclei not occurring in nature might be stable enough to be prepared by nuclear reactions.

Another model, which is particularly useful for explaining nuclear fission, is the "liquid drop" model. Whereas the shell model pictures the nucleus as a highly organized structure, the liquid drop model assumes a random arrangement of nuclear particles, analogous to the arrangement of the molecules in a drop of liquid. Such a nucleus can be easily deformed. Thus, if the nucleus is given enough energy, the deformation can be severe enough to cause the nucleus to fission into two or more pieces.

Still a third model, the "collective model," combines the more successful features of the shell model with those of the liquid drop model.

Suggestions for further reading

Baranger, M., and Sorenson, R. A., "The Size and Shape of Atomic Nuclei," *Scientific American,* August 1969, p. 58.

Choppin, G. R., *Nuclei and Radioactivity* (New York: W. A. Benjamin Co., Inc., 1964).

Curie, E., *Madame Curie* (New York: Garden City Publishing Co., 1939).

Feinberg, G., "Ordinary Matter," *Scientific American,* 1967, p. 126.

Hyde, E. K., "Nuclear Models," *Chemistry, 40,* July 1967, p. 12.

Mayer, Maria G., "The Structure of the Nucleus," *Scientific American,* March 1951, p. 22.

Peierls, R. E., "Models of the Nucleus," *Scientific American,* January 1959, p. 75.

Yagoda, H. "The Tracks of Nuclear Particles," *Scientific American,* May 1956, p. 40.

Questions

1. Assuming that nuclei consist of protons and neutrons, describe the following transformations:
(*a*) The reaction of a nitrogen nucleus with an alpha particle to produce an oxygen nucleus and a proton, discovered by Rutherford (p. 308).
(*b*) The reaction of a beryllium-9 nucleus with an alpha particle to produce a carbon-12 nucleus and a neutron, discovered by Chadwick (p. 309).

2. The proton was discovered some twelve years before the neutron. During this period, a theory of the nucleus was developed in which the nucleus was thought to consist of protons and *electrons*. Can you suggest what this theory might have been like?

3. Could Rutherford's scattering experiment have been carried out successfully using neutrons instead of alpha particles? Explain.

4. Construct a table which gives the number of protons, the number of neutrons and the number of extra-nuclear electrons for the following atoms: C^{13}, C^{14}, O^{17}, F^{19}, Zn^{66}, U^{237}, Pb^{207}.

5. Write nuclear symbols for the ten isotopes of tin, whose mass numbers are 112, 114, 115, 116, 117, 118, 119, 120, 122 and 124.

6. How does the "mass number" differ from the atomic weight of an element?

7. Draw a schematic cloud track for a reaction in which an alpha particle reacts with a beryllium-9 nucleus to produce a carbon-12 nucleus and a neutron.

8. Why is the Geiger counter a rather poor device for detecting neutrons?

9. (*a*) What is the evidence that the forces binding neutrons to neutrons in the nucleus are not electrostatic?
(*b*) If the helium nucleus contains two positively charged protons that repel each other, why doesn't it fly apart?

10. List five nuclei that you would expect to be especially stable (that is, non-radioactive), and explain why.

When experiment is pushed into new domains, we must be prepared for new facts, of an entirely different character from those of our former experience.

PERCY W. BRIDGMAN (1927)

Eighteen

NUCLEAR REACTIONS AND NUCLEAR ENERGY

Radioactive decay as a nuclear reaction

NATURAL RADIOACTIVITY was first mentioned in Ch. 8 as evidence that the one-piece, indestructible atom envisioned by John Dalton was unsatisfactory. It will be recalled that radioactive decay involves the spontaneous disintegration of the nuclei of certain isotopes, with the emission of an alpha particle or a beta particle and the production of a nucleus of a different element. For example, atoms of thorium-232 are converted spontaneously to atoms of radium-228, with each conversion accompanied by the emission of an alpha particle ($_2He^4$). The radium-228 produced from the decay of thorium-232 is also radioactive and decays with the emission of a beta particle ($_{-1}e^0$) to form radioactive actinium-228. Disintegration continues through a series of steps until, finally, all of the thorium-232 is converted to lead-208. The process takes a very long time, with half the thorium decaying in a matter of some ten billion years.

There are over 40 naturally occurring radioactive isotopes, most of them of high atomic weight. A few of the more important ones are listed in the table on the following page.

The half-life of a radioactive isotope is a property characteristic of that particular isotope and is equal to the time it takes for half of the original sample to decay. Thus, if an isotope has a half-life of 10 years, half of the original sample will have undergone disintegration at the end of 10 years. After an additional 10 years, or a total of 20 years, half of the remaining half will have decayed, so that a quarter of the original is left ($\frac{1}{2} \times \frac{1}{2} = \frac{1}{4}$). Similarly, after 30 years or three half-lives, one eighth of the original sample remains ($\frac{1}{2} \times \frac{1}{2} \times \frac{1}{2} = \frac{1}{8}$); and so on.

Note that the decay rate is always based on the amount of material *left*. It is important to avoid the erroneous assumption that the entire sample of material will decay in just twice the half-life. The right and the wrong views of radioactive decay are contrasted in Fig. 18-1.

The range of half-lives for different radioactive isotopes is enormous, from less than a ten-millionth (10^{-7}) of a second to billions of billions (10^{18}) of years. Under ordinary conditions, the half-lives are independent of the temperature of the sample and of the nature of the *host molecule,* that is, the molecule of which the radioactive atom is a part. This property of a fixed and characteristic half-life has been of great usefulness in determining the age of materials containing radioactive atoms. Our best estimates of the age of the earth are based on this concept. Knowing the half-life of a given radioactive isotope, we need only determine the ratio of parent atoms to decay products in the sample. It is then a simple matter to calculate the amount of time required to reach this ratio, assuming that at the beginning of the process only

SOME NATURALLY-OCCURRING RADIOACTIVE ISOTOPES

Isotope	Particle emitted	Half-life (years)
C^{14}	Beta	5.7×10^{3}
K^{40}	Beta	1.4×10^{9}
Rb^{87}	Beta	6×10^{10}
In^{115}	Beta	6×10^{14}
Sn^{124}	Beta	Greater than 1.7×10^{17}
La^{138}	Beta	1.2×10^{12}
Re^{187}	Beta	4×10^{12}
Ra^{226}	Alpha	1.6×10^{3}
Th^{232}	Alpha	1.4×10^{10}
Th^{230}	Alpha	8.0×10^{4}
Pa^{231}	Alpha	3.4×10^{4}
U^{235}	Alpha	7.1×10^{8}
U^{238}	Alpha	4.5×10^{9}

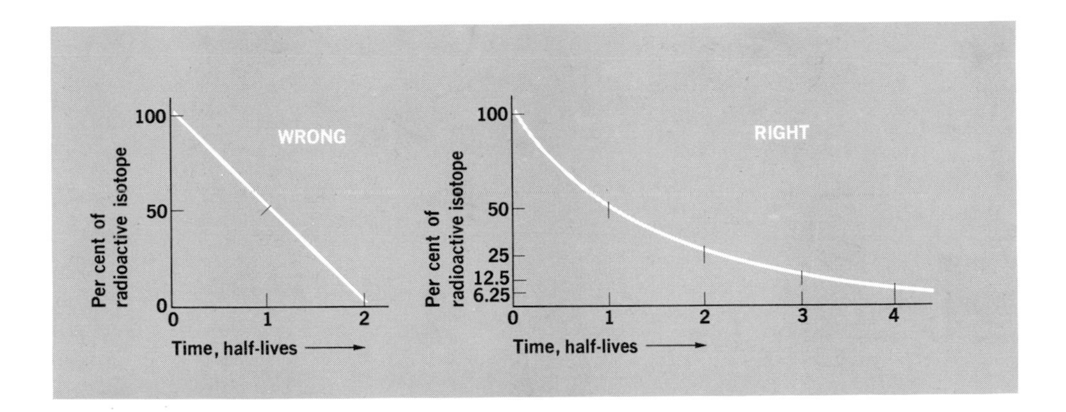

Fig. 18-1
Graphs of the wrong and the right conception of radioactive decay.

parent atoms were present. On this basis, the age of the oldest rocks in the earth's crust has been estimated as four billion years.

ALPHA EMISSION

Let us examine the process of radioactive decay in terms of the proton-neutron theory of the nucleus. The first example to be considered is that of an alpha emitter, that is, of a nucleus which decays by the emission of an alpha particle. There are two major questions to be considered: first, what happens to the parent nucleus as a consequence of this emission; and second, why do certain isotopes undergo this type of spontaneous disintegration while others do not?

One can best understand the consequences of alpha emission by considering a few examples, keeping a close audit on the various nuclear particles involved in the process.

Uranium-238 ($_{92}U^{238}$) is an alpha emitter. The nucleus of $_{92}U^{238}$ consists of 92 protons and 146 neutrons. The alpha particle ($_2He^4$) consists of 2 protons and 2 neutrons. The ejection of an alpha particle is therefore equivalent to the loss of 2 protons and 2 neutrons by the uranium-238 nucleus. As a consequence, the daughter nucleus consists of 90 protons and 144 neutrons. But such a nucleus is no longer a nucleus of uranium; it is a nucleus of the element thorium, with atomic number 90 and atomic mass 234. These facts are conveniently summarized as follows:

92 p 146 n	$\longrightarrow$	2 p 2 n	+	90 p 144 n
Uranium-238 nucleus		Alpha particle		Thorium-234 nucleus

or even more compactly:

$$_{92}U^{238} \longrightarrow {}_{2}He^{4} + {}_{90}Th^{234}$$

Since this is an equation, the amounts involved on both sides of the arrow must balance, that is, the atomic number of the parent must equal the sum of the atomic numbers of the products, and similarly for the mass numbers.

Thorium-234 is also an alpha emitter. Since thorium-234 contains 90 protons, the ejection of an alpha particle will reduce the number of protons to 88. The loss of two neutrons will, in a similar fashion, reduce the number of neutrons from 144 to 142. The result is the production of a daughter nucleus with a mass number that is four less than that of the parent nucleus, and with a nuclear charge (or atomic number) that is two less, namely the nucleus of a radium atom:

90 p 144 n	$\longrightarrow$	2 p 2 n	+	88 p 142 n
Thorium-234 nucleus		Alpha particle		Radium-230 nucleus

or

$$_{90}Th^{234} \longrightarrow {}_{2}He^{4} + {}_{88}Ra^{230}$$

EMISSION OF NEGATIVE BETA PARTICLES

The case of beta emission is somewhat more complicated, but it is still understandable in terms of the proton-neutron theory. Two kinds of beta particles are known, differing only in the sign of their charge. The negative beta particle is the same as an electron, that is, it has a mass of zero on the atomic mass scale and bears a single negative charge. It can therefore be symbolized as $_{-1}e^0$. The positive beta particle is called a positron and will be discussed later.

Electrons *per se* are not present in the nucleus. The fact that they emerge from the nucleus can only mean that they are somehow generated in the nucleus and then ejected immediately. What sort of process could lead to the formation of an electron within the nucleus? A readily envisioned process is the conversion of a neutron ($_0n^1$) to a proton ($_1H^1$) plus an electron, as illustrated in the following equation:

$$_0n^1 \longrightarrow {}_{-1}e^0 + {}_1H^1$$

The charges of the newly formed particles (-1 and $+1$) add up to zero, and the masses (0 and 1) add up to one. Since these quantities are equal to the charge and mass, respectively, of the neutron, our postulate is satisfactory, at least on a "bookkeeping basis."

Now, if the emission of a negative beta particle involves the conversion of a neutron to a proton, then the mass number of the beta-emitting nucleus will not change, but the atomic number (equal to the number of protons in the nucleus) will increase by one. Consider, for example, the case of carbon-14, which is a beta emitter. When a carbon-14 nucleus decays, the daughter nucleus is that of nitrogen-14:

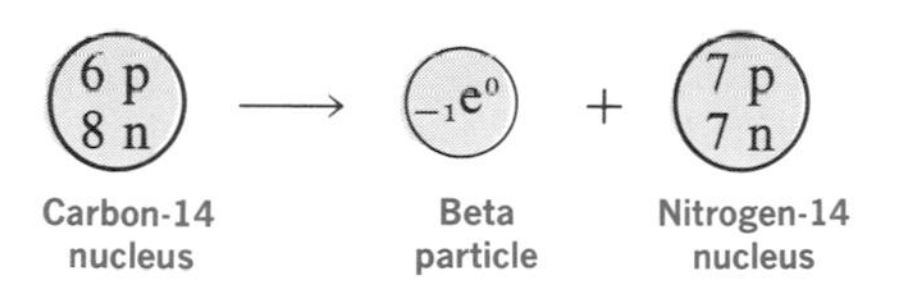

Examination of this schematic representation of the decay process shows that one of the 8 neutrons in the C-14 nucleus has been converted to a proton, giving 7 protons and 7 neutrons in the daughter nucleus, which is therefore nitrogen-14. More briefly, this nuclear change can be represented as:

$$_6C^{14} \longrightarrow {}_{-1}e^0 + {}_7N^{14}$$

In this equation, both the mass numbers and the nuclear charges are balanced, following the same rules as were stated for the case of alpha emission.

DETERMINATION OF AVOGADRO'S NUMBER

The development of techniques for the counting of individual alpha particles has provided us with another tool for the direct determination of Avogadro's number, that is, the number of molecules in one gram molecular weight or the number of atoms in one gram atomic weight. Since alpha particles are nuclei of helium atoms, they need only acquire a pair of electrons to become ordinary helium atoms. This happens readily after the alpha particles have slowed down.

One gram of radium is known to emit 4.35×10^{18} alpha particles per year. The British scientists, Ramsay and Soddy, had shown that a gram of radium in one year produces enough alpha particles to form 0.158 cc of helium gas (under standard conditions), or 2.82×10^{-5} g of helium. Thus, 2.82×10^{-5} g of helium consists of 4.35×10^{18} helium atoms. It follows that one gram atomic weight of helium, weighing 4.00 g, consists of $4.00 \times 4.35 \times 10^{18} \div 2.82 \times 10^{-5}$, or 6.16×10^{23} atoms. This value is very close to the most accurate current value for Avogadro's number, 6.023×10^{23} atoms per gram atomic weight.

Artificially induced nuclear reactions

So far we have discussed just one type of nuclear reaction: that which occurs during the disintegration of a naturally radioactive substance. In addition to this type, there are a wide variety of other nuclear reactions that can be brought about by the *bombardment* of nuclei with such subatomic particles as protons, alpha particles, neutrons, and electrons, or with gamma rays. These include

1. transmutation to a new stable isotope,
2. transmutation to a new radioactive isotope,
3. nuclear fission, and
4. nuclear fusion.

Which of these several reactions occurs depends on

1. the kind of nucleus being bombarded,
2. the nature of the bombarding particle, and
3. the *energy* of the bombarding particle.

In the following sections, these various factors will be considered in turn.

In order for an artificially induced nuclear reaction to occur, the target nucleus must be bombarded by some subatomic particle. In all cases except those involving neutrons, electrons, or gamma rays, the bombarding particle has a positive charge. Since the nuclei of atoms are also positively charged, it is obvious that there will be an electrostatic repulsion to be overcome before the bombarding particle (projectile) can enter the target nucleus. This repulsion is overcome if the projectile is given a high velocity and, therefore, considerable kinetic energy.

In the earliest experiments involving nuclear transformation, the only projectiles available were those ejected during the course of the disintegration of naturally radioactive atoms. Thus, Rutherford used the alpha particles ejected by polonium as projectiles in the experiment which resulted in the transformation of nitrogen atoms into oxygen atoms. The limitation inherent in the use of naturally available particles is that their energy is determined by the nature of the radioactive decay process and cannot be varied at will. To overcome this limitation and to have available particles of very high kinetic energy, various devices such as the cyclotron (Fig. 18-2), the Van de Graaff accelerator, and the betatron have been invented. The operating principle of all accelerators is the same, although the details of construction and operation vary widely, depending on the energy range and the type of particle employed. In every case, because the particle is charged, it can be accelerated if one takes advantage of the attractive and repulsive forces that exist between unlike and like charges, respectively.

These attractive and repulsive forces serve to push the particles to high velocities so that, on encountering the target nucleus, they are able to produce the desired nuclear reaction.

Nuclear transmutation

Using either the naturally available alpha particles, or artificially accelerated alphas, protons, electrons, and so forth, it is possible for one to carry out a large variety of nuclear reactions leading to both stable and artificially radioactive products. In Eqs. (1) to (4), a number of examples of nuclear transformations are given:

$$_{7}N^{14} + _{2}He^{4} \longrightarrow _{8}O^{17} + _{1}H^{1} \quad (1)$$
$$_{13}Al^{27} + _{2}He^{4} \longrightarrow _{15}P^{30} + _{0}n^{1} \quad (2)$$
$$_{11}Na^{23} + _{1}H^{1} \longrightarrow _{12}Mg^{23} + _{0}n^{1} \quad (3)$$
$$_{3}Li^{7} + _{1}H^{2} \longrightarrow _{3}Li^{8} + _{1}H^{1} \quad (4)$$

Equation (1) represents the bombardment of nitrogen-14 atoms with alpha particles and is, of course, the classical experiment of Rutherford. Equation (2) represents the first transmutation leading to the production of a radioactive isotope not found in nature. The reaction was first carried out in 1934 by Irene Joliot-Curie, the daughter of the discoverers of radium and polonium, and her husband, Frédèric Joliot.

Phosphorus-30, the product in Eq. (2), decays by ejecting a *positron* or positively charged electron, $_{+1}e^{0}$, as shown in Eq. (5).

$$_{15}P^{30} \longrightarrow _{14}Si^{30} + _{+1}e^{0} \quad (5)$$

Equation (3) depicts the transmutation of sodium-23 into magnesium-23 by proton bombardment, and the net result of Eq. (4) is the addition of a neutron to the lithium nucleus to produce the isotope with the next higher mass number.

A useful theory concerning the *mechanism* of nuclear reactions assumes that they take place in two steps. In the first step, the incident particle is absorbed by the target nucleus to form a *compound nucleus*. This compound nucleus is, of course, very unstable and immediately breaks down into the final products. Using this theory, we may depict the mechanism of reaction (1) as follows:

$$\underset{\text{Projectile}}{_{2}He^{4}} + \underset{\text{Target}}{_{7}N^{14}} \longrightarrow \underset{\text{Unstable compound nucleus}}{[_{9}F^{18}]} \longrightarrow \underset{\text{Products}}{_{8}O^{17} + _{1}H^{1}}$$

Similarly, the mechanism of reaction (3) may be depicted as:

$$\underset{\text{Projectile}}{_{1}H^{1}} + \underset{\text{Target}}{_{11}Na^{23}} \longrightarrow \underset{\text{Unstable compound nucleus}}{[_{12}Mg^{24}]} \longrightarrow \underset{\text{Products}}{_{12}Mg^{23} + _{0}n^{1}}$$

In certain nuclear transmutations, the nucleus is left in a state of high energy, in which case this excess energy may be disposed of by the emission of a gamma ray. This is strongly reminiscent of the situation involving the orbital electrons of excited or energetic atoms which also rid themselves of their excess energy by the emission of radiation (see pp. 148–156).

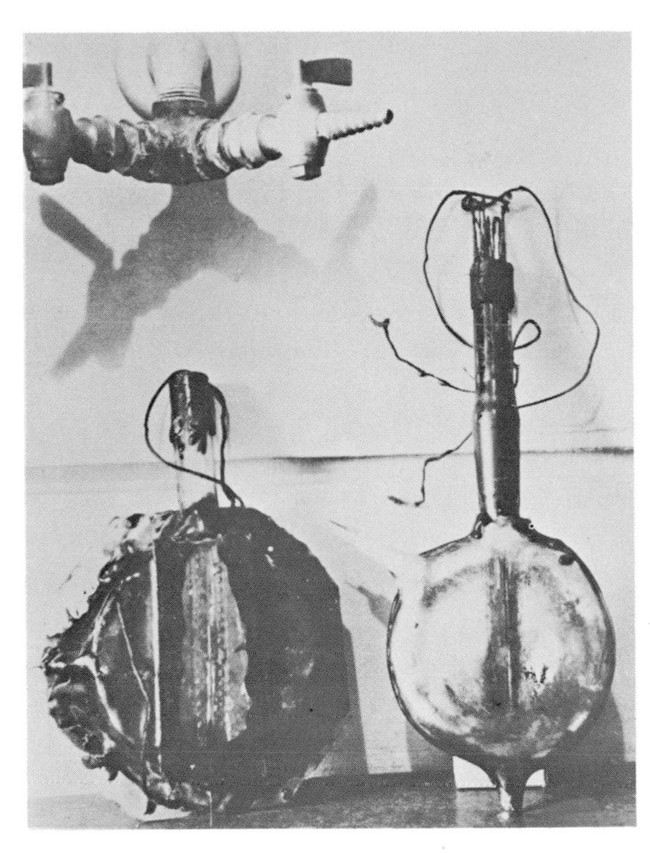

Fig. 18-2

The vacuum chambers from the first two cyclotrons, (top left) built by E. O. Lawrence and N. E. Edlefsen in 1930. They are about six inches in diameter. The lower picture shows the 184-inch cyclotron at the Radiation Laboratory, Berkeley, California.

Nuclear energy

A sample of a highly radioactive substance is capable of heating itself to a temperature appreciably above that of its surroundings. If the isotope is long-lived, it will continue to do this for a very long time. Thus, a gram of radium-226, over a period of one year, will produce 1,000,000 calories of heat, and will continue to produce heat at a rate that diminishes only slowly for many centuries. What is the source of all this energy?

Again, if we compare the kinetic energy of the projectile and of the target nucleus with the kinetic energy of the *product nuclei,* we find that the kinetic energy of the products is frequently much greater than that of the starting materials. For example, in the reaction,

$${}_3Li^7 + {}_1H^1 \longrightarrow {}_2He^4 + {}_2He^4$$

the proton must have a kinetic energy of 200,000 electron volts in order that the reaction may take place when the lithium atom is at rest (zero kinetic energy).† After the reaction has taken place, it is found that *each* of the ${}_2He^4$ nuclei has a kinetic energy of 8,700,000 ev, for a total of 17,400,000ev. Subtracting the initial 200,000ev, we see that 17,200,-000ev of energy is produced for every lithium atom destroyed.

It is of interest to compare this amount of energy with the amounts of energy liberated in a typical *chemical* reaction. Thus, if water is formed by the combination of hydrogen and oxygen, the energy liberated for every molecule of water produced is only 3.0 ev, while in the nuclear reaction involving ${}_3Li^7$, 5,700,000 times as much energy is liberated per atom of lithium. It is generally the case that, whenever energy is liberated in a nuclear reaction, the amount liberated per nucleus reacted is on the order of millions of times greater than the amount liberated per molecule reacted in an ordinary chemical reaction. Again we may ask, what is the source of all this energy?

The answer to this question is to be found by a careful examination of the *masses* of particles involved in the nuclear reactions. To this point

† The *electron volt* (ev) is a convenient unit for expressing the energy involved in nuclear reactions. It is the amount of energy acquired by an electron when accelerated through a potential difference of one volt. One electron volt per particle is equivalent to 23,053 calories per mole of particles.

in the discussion, the masses of the nuclear particles and of the nuclei themselves have been rounded off to the nearest whole number, which was called the mass number. In point of fact, the masses of the proton, neutron, and the various isotopes of each of the elements are known with considerable accuracy, and they are not integral numbers. When we consider these accurate masses, a very startling fact becomes apparent. For example, while the mass numbers in the reaction

$${}_1H^1 + {}_3Li^7 \longrightarrow {}_2He^4 + {}_2He^4 + \text{Kinetic energy}$$

add up to *eight* on both sides of the equation, a discrepancy is noticed when the *accurate masses* are used.

	${}_1H^1$	+ ${}_3Li^7$	$\longrightarrow$ ${}_2He^4$	+ ${}_2He^4$
Accurate mass:	1.00815	7.01822	4.00388	4.00388

The masses on the left add up to 8.02637, while on the right, the masses of the two helium nuclei add up to 8.00776. Thus, there is a discrepancy of 0.01861 mass units. In other words, during the course of this nuclear transformation, 0.01861 units of mass have disappeared, in violation of the law of conservation of mass! The present example is not an isolated one. In every case where energy is released in a nuclear reaction, it is found that the mass of the products is somewhat less than the mass of the reactants. It would seem that the law of conservation of mass does not apply to nuclear reactions.

Long before these experimental facts were known, Albert Einstein had supplied us with a possible explanation. In connection with his special theory of relativity, he had come to the conclusion that mass and energy are not independent entities, but that the mass of a body is actually a measure of its energy. Thus, when a body loses energy, its mass must decrease by a proportionate amount. The relationship that Einstein suggested in 1905 is given by the now well-known equation,

$$E = mc^2$$

where E represents the quantity of energy, m the amount of mass equivalent to that energy, and c^2 is the square of the velocity of light. Since m is multiplied by c^2, which is a *very* large number (expressed in familiar units, the velocity of light is 186,000 mi/sec), it is clear that even a very tiny amount of matter is equivalent to a very large amount of energy.

Albert Einstein: 1879–1955
[Photograph courtesy Brown Brothers.]

Using this equation, we can calculate that the complete conversion of one atomic mass unit of matter (1.66×10^{-24} g) will result in the release of 931 million electron volts (Mev) of energy. Employing this relationship, let us see if the mass discrepancy in the $_3Li^7$ reaction previously discussed can be accounted for in terms of the kinetic energy liberated in the reaction:

Let $X =$ the energy equivalent to 0.01861 mass units, the mass discrepancy in this reaction. Since 931 Mev is equivalent to 1 mass unit, by simple proportion we have that

$$\frac{931 \text{ Mev}}{1 \text{ Mass unit}} = \frac{X \text{ Mev}}{0.01861 \text{ Mass units}}$$

Solving for X, we find that:

$$X = 931 \times 0.01861$$
$$X = 17.3 \text{ Mev}$$

This answer very nearly equals the 17.2 Mev of kinetic energy observed experimentally. Such good agreement would seem to confirm the correctness of Einstein's equation.

A cluster of protons and neutrons is a stable nucleus only if the total energy of the cluster is less than the total energy of the separated particles. In view of Einstein's equation, this means that the mass of the nucleus must be less than that of the separated protons and neutrons of which it is composed. This is always the case. Consider, for example, the formation of a helium nucleus from two protons and two neutrons:

	$2\,{}_1H^1$	$+$	$2\,{}_0n^1$	$\longrightarrow$	${}_2He^4$
Accurate mass:	2×1.00815		2×1.00899		4.00388

The combined mass of the two protons and two neutrons is 4.03428, whereas the mass of the helium nucleus is only 4.00388. Thus, there is a decrease in mass and, hence, of energy, equivalent to 0.03040 mass units, or 28.3 Mev.

This energy, which is given up when the helium nucleus is formed, is called the *binding energy* since it is a measure of the tenacity with which the nuclear particles adhere to one another in the nucleus. Thus, if we wish to separate a helium nucleus into two protons and two neutrons, we must supply exactly this amount of energy.

It is of interest to calculate the average binding energy *per nuclear particle,* that is, the total binding energy divided by the total number of protons and neutrons. For helium, this quantity is $28.3 \div 4$, or 7.08 Mev. For other nuclei, the average binding energy per nuclear particle may of course be different. For example, for bismuth-209, the average binding energy is found to be 7.8 Mev per nuclear particle, indicating that the nuclear particles are bound more firmly in this nucleus than in the helium nucleus. A plot of the average binding energy per nuclear particle for the various stable isotopes is given in Fig. 18-3. Examination of this figure shows that the most stable nuclei, that is, those with the greatest binding energy per particle, are to be found in the mass number range 40–120 (corresponding to the elements of atomic numbers 20 to 50).

Figure 18-3 provides some insight into why the elements of high atomic weight are radioactive, and particularly, why they usually decay by alpha emission. It can be seen from the figure that the binding energy per nuclear particle is relatively low for the very heavy elements. Thus, any process that converts a heavy nucleus to one of intermediate atomic weight, and consequently, greater binding energy, will be favored. Such a conversion is accomplished by the loss of an alpha particle, which lowers the atomic mass by four units. It is usually an alpha particle that

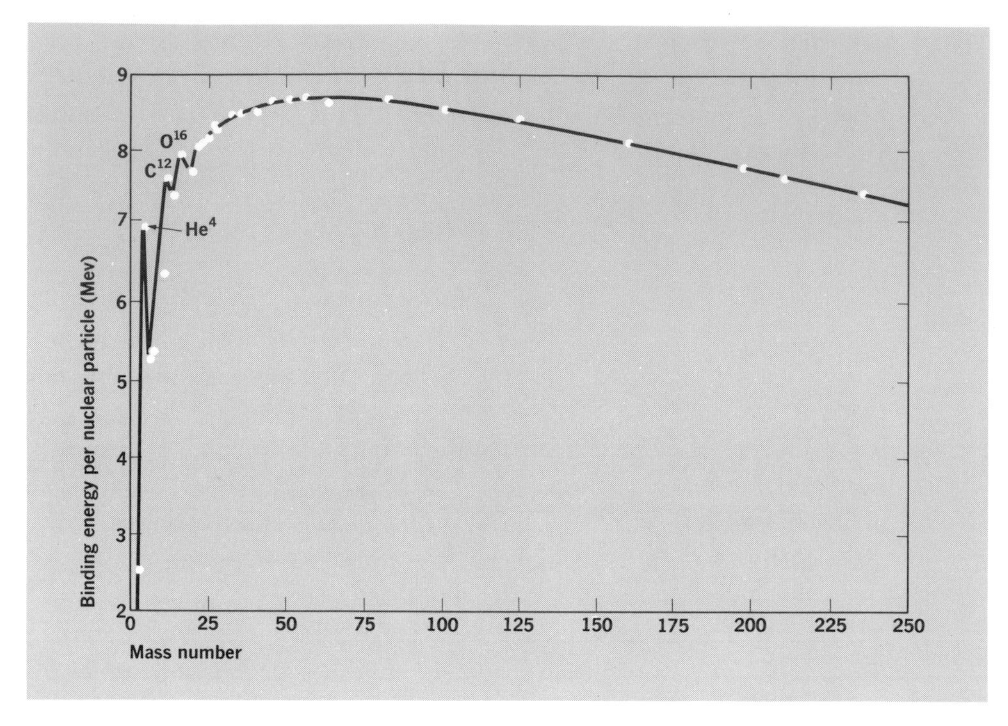

Fig. 18-3
Plot of the average nuclear binding energy per particle as a function of nuclear mass number. [Courtesy of S. Glasstone, *Source Book on Atomic Energy*, 2nd Ed. (Princeton, N. J.: D. Van Nostrand, 1958), p. 351.]

is lost simply because this is an especially stable light particle, as can also be seen from Fig. 18-3.

In view of the large amounts of energy released in the nuclear reactions discussed in this section, one might think that these reactions could be employed for the commercial production of nuclear energy. However, there are two reasons why they are not. First, one must expend large amounts of energy in accelerating the projectiles to the required velocity; and second, in view of the small size of the nucleus, only a small fraction of the accelerated projectiles actually collide with a nucleus and react. As a result, the energy used up in accelerating a large number of particles greatly exceeds the energy released in the relatively few reactions which occur.

Nuclear fission

There is, however, a nuclear reaction which does not require that the bombarding particles be accelerated, and which can be made to be a prolific source of energy. As a consequence of earlier work by the Italian physicist, Enrico Fermi, directed toward the production of elements with atomic numbers greater than that of uranium, the German scientists, Otto Hahn and Fritz Strassmann, in 1938, bombarded samples of uranium with neutrons. In addition to the hoped-for heavier elements, they discovered that their sample, after bombardment, also contained atoms of the lighter element, barium, and of other elements with atomic numbers considerably *less* than that of uranium. At the time of their experiments, there was no known nuclear reaction mechanism that could account for the production of these nuclei from uranium. However, soon thereafter (1939), two other Germans, Lise Meitner and Otto Frisch, offered an explanation for these observations which is quite consistent with the experimental facts. They suggested that, instead of the usual small fragments being ejected during the nuclear reaction, the uranium nucleus was split in two by the incident neutron. Thus, they envisioned that such elements as barium and krypton might be produced by the following mechanism:

$${}_0n^1 + {}_{92}U^{235} \longrightarrow \underset{\text{Unstable compound nucleus}}{[{}_{92}U^{236}]}$$

$$[{}_{92}U^{236}] \longrightarrow {}_{56}Ba^{143} + {}_{36}Kr^{90} + 3\,{}_0n^1$$

It is now known that many different isotopes, ranging in mass number from 72 to 160, are produced in the fission process. Some other examples of the many different ways in which the uranium nucleus may split are shown in the following equations:

$$[{}_{92}U^{236}] \longrightarrow {}_{54}Xe^{139} + {}_{38}Sr^{95} + 2\,{}_0n^1$$

$$[{}_{92}U^{236}] \longrightarrow {}_{53}I^{135} + {}_{39}Y^{97} + 4\,{}_0n^1$$

Enrico Fermi: 1901–1954
Fermi is noted for his theory of beta decay and for his experimental studies of neutron-induced nuclear reactions. He directed the construction of the first nuclear chain reactor. [Photograph courtesy Brown Brothers.]

Examination of these equations shows that, in addition to the two principal fragments, each fission leads to the formation of several neutrons.

The energy released in such a process is about 200 Mev, or more than ten times as much as is released in typical nuclear reactions not involving fission. The release of such a large amount of energy is, of course, accompanied by a corresponding decrease in mass. The total mass of fission products plus neutrons is about 0.215 atomic mass units less than the mass of the uranium atom plus the incident neutron.

Consideration of Fig. 18-3 provides an explanation for this nuclear breakdown with the release of large amounts of energy. The very heavy nuclei have a lower average binding energy than those of intermediate mass number. Therefore, the heavy nuclei will, if possible, undergo conversion to the more stable intermediate elements with the release of energy. This energy is in the form of kinetic energy of the fission fragments, plus some radioactive decay energy, since the fission fragments themselves are usually radioactive. There is also considerable gamma radiation, which is released at the instant the fission takes place. Eventually, most of this energy is converted to heat.

Since each fission requires only one neutron but produces several (2.5 neutrons on the average), the possibility exists that these product neutrons can go on to cause the fission of additional uranium-235 nuclei. When this happens, a branching chain reaction results, as shown schematically in Fig. 18-4.

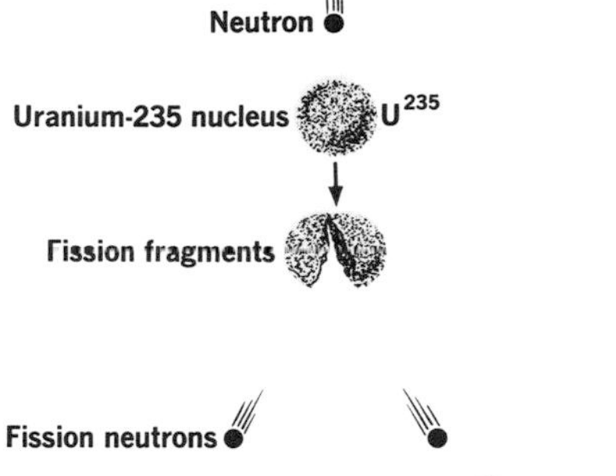

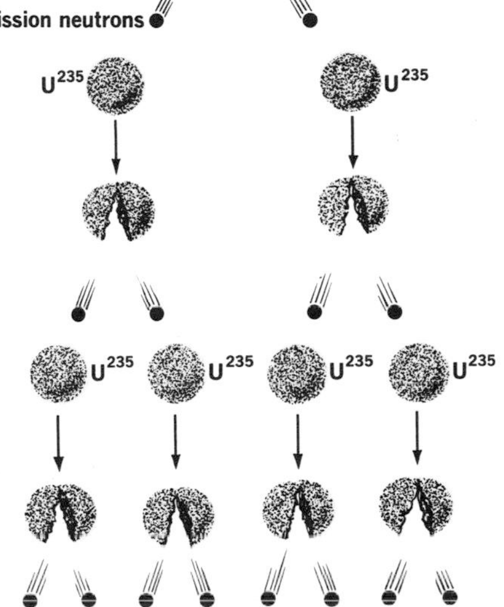

Fig 18-4
Schematic representation of a branching nuclear chain reaction.

It is now history, of course, that such a chain reaction was first achieved in 1942 by a group of scientists at the University of Chicago, working under the direction of Enrico Fermi. They constructed a nuclear reactor consisting of lumps of uranium and uranium oxide embedded in a lattice of graphite. Under these conditions, the rate of the fission process can be controlled, and energy is released at a rate slow enough to permit the use of the reactor as a source of power. In the so-called "atomic" bomb (fission bomb), a large number of fissions occur *very* rapidly with an explosive release of energy.

Nuclear fusion

An examination of Fig. 18-3 reveals that the elements of *low* atomic number are also less stable than those of intermediate atomic number, just as those of high atomic number are unstable with respect to the intermediate group of elements. We have seen that, under certain circumstances, the very heavy elements can be split to produce elements of intermediate weight with the accompanying release of prodigious quantities of energy. In a similar fashion, we would expect that it might be possible to *fuse* the nuclei of some of the lighter elements to form intermediate elements of greater stability, and that this process, too, would lead to the release of large amounts of energy. It has long been postulated that fusion reactions of just this type are the source of the sun's energy. In 1952, the fusion reaction was achieved by artificial means in what is commonly called the "hydrogen bomb." In this particular device, an ordinary fission bomb is used to raise the temperature of the reacting nuclei to millions of degrees, at which temperature they have sufficient kinetic energy to fuse when they collide. The fusion reaction releases energy at the rate of from 3 to 20 Mev per fusion event. A number of examples of fusion reactions follow:

$${}_1H^2 + {}_1H^2 \longrightarrow {}_2He^3 + {}_0n^1$$
$${}_1H^2 + {}_1H^2 \longrightarrow {}_1H^3 + {}_1H^1$$
$${}_1H^3 + {}_1H^2 \longrightarrow {}_2He^4 + {}_0n^1$$

Since there are estimated to be some 10^{17} pounds of deuterium (${}_1H^2$) in the waters of the earth, and since each pound is equivalent in energy to about 2500 *tons* of coal, it is quickly appreciated that a controlled fusion reactor would provide a virtually inexhaustible supply of energy. A great effort is being made today to build such a reactor, and there is every reason to hope that success may not be too far in the future.

Suggestions for further reading

Badash, L., "How the Newer Alchemy Was Received," *Scientific American,* August 1966, p. 88.

Choppin, G. R., "Nuclear Fission," *Chemistry, 40,* July 1967, p. 25.

Clark, M., "The Origin of Nuclear Science," *Chemistry, 40,* July 1967, p. 8.

Fermi, L., *Atoms in the Family* (Chicago: University of Chicago Press, 1953).

Hahn, O., "The Discovery of Fission," *Scientific American,* February 1958, p. 76.

Hudis, J., "Nuclear Reactions," *Chemistry, 40,* July 1967, p. 20.

Johnsen, R. H., "Radiation Chemistry," *Chemistry, 40,* July 1967, p. 31.

Levinger, E. E., *Albert Einstein* (New York: J. Messmer, Inc., 1949).

Reynolds, J. H., "The Age of the Elements in the Solar System," *Scientific American,* November 1960, p. 171.

Seaborg, G. T., "The Synthetic Elements," *Scientific American,* April 1969, p. 56.

Teeter, C. E., "An Introduction to Nuclear Power," *Journal of Chemical Education, 47,* March 1970, p. 208.

Wahl, W. H., and Kramer, H. H., "Neutron-Activation Analysis," *Scientific American,* April 1967, p. 68.

Wolfgang, R., "Chemical Accelerators," *Scientific American,* October 1968, p. 44.

Questions

1. (*a*) Potassium-40 is a naturally occurring beta emitter with a half-life of 1.4×10^9 years. If the earth is 4.2 billion years old, what fraction of the potassium-40 present initially has decayed by now?
(*b*) What is the product of radioactive decay of potassium-40?

In questions 2–8, complete the nuclear equations.

2. ${}_{8}O^{15} \longrightarrow {}_{7}N^{15} + \underline{\quad ? \quad}$

3. $\underline{\quad ? \quad} \longrightarrow {}_{+1}e^{0} + {}_{37}Rb^{85}$

4. ${}_{33}As^{76} \longrightarrow {}_{-1}e^{0} + \underline{\quad ? \quad}$

5. ${}_{8}O^{16} + \underline{\quad ? \quad} \longrightarrow {}_{8}O^{17} + {}_{1}H^{1}$

6. ${}_{92}U^{238} + \underline{\quad ? \quad} \longrightarrow {}_{92}U^{239}$

7. ${}_{83}Bi^{209} + \underline{\quad ? \quad} \longrightarrow {}_{85}At^{211} + 2{}_{0}n^{1}$

8. $\underline{\quad ? \quad} + {}_{1}H^{2} \longrightarrow {}_{0}n^{1} + {}_{5}B^{10}$

9. Calculate the amount of energy that would be liberated if 0.5 g of helium were completely converted into energy. How much energy would be liberated if 0.5 g of iron were converted into energy instead of 0.5 g of helium?

10. Calculate the amount of energy that is released when three protons and four neutrons combine to form ${}_{3}Li^{7}$. The accurate mass of lithium-7 is 7.01822.

11. Which type of nuclear reaction, fission or fusion, produces the most energy per atom of material reacted? (Examine Fig. 18-3.)

12. Explain why, in order for a branching chain reaction to take place, there must be more than one neutron released per fission event.

Appendix A

SOME SCIENTIFIC UNITS

MASS The standard unit of mass is the *kilogram* (kg), roughly 2.2 pounds. The *international prototype kilogram* is a metal cylinder fabricated from a corrosion-resistant platinum-iridium alloy and is preserved at the International Bureau of Weights and Measures in Sèvres, France. One gram (g) is equal to one-thousandth of a kilogram.

WEIGHT The weight of an object is a measure of how strongly the object is attracted by the earth. The unit of weight has been defined so that, at mean sea level, an object with a mass of one gram also has a weight of one gram. At other points on the earth's surface, the mass (in grams) and the weight (in grams) differ slightly, but the difference is so small that, for most practical purposes, the terms mass and weight may be used interchangeably.

LENGTH The standard unit of length is the *meter* (m), equivalent to 1.0936 yards or 39.37 inches. The *international prototype meter* is the distance between two lines engraved on a platinum-iridium bar, kept at 0 °C, currently on deposit at the International Bureau of Weights and Measures. The original intention was to define one meter as 1/40,000,000 of the mean circumference of the earth, but owing to error, the international prototype meter is actually closer to 1/40,009,000 of the earth's mean circumference.

A convenient derived unit is the *centimeter* (cm) which is equal to one-hundredth of a meter. One centimeter is close to four-tenths of an inch.

VOLUME The unit of volume in the metric system is the *cubic centimeter* (cc), which is the volume of a cube 1 cm on each side. One cubic centimeter is equal to 0.0010567 quart, or roughly one-thousandth of a quart.

Another unit of volume employed in scientific work is the *liter* (l), which is the volume occupied by 1 kg of water at 4 °C. A derived unit is the *milliliter* (ml), or one-thousandth of a liter. One milliliter is so nearly equal to one cubic centimeter (1 ml = 1.000027 cc) that the two units are almost always used interchangeably.

TIME The primary standard in the measurement of time is the mean solar day, which consists of 24 hours or 86,400 seconds. In scientific work, it is customary to express time intervals in *seconds* (sec).

TEMPERATURE In scientific work, the centigrade scale of temperature is used. On this scale, the melting point of ice is defined as 0°, and the boiling point of water at a pressure of 1 atm as 100°. The interval between melting and boiling points is thereby divided into one hundred equal units, each called one degree centigrade (1 °C). In practice, temperatures are measured by means of a substance, some convenient property of which varies with the temperature. The most commonly used thermometer is made of liquid mercury sealed in a glass tube. An increase in temperature causes an elongation of the mercury column.

The familiar household thermometer employs the Fahrenheit scale. On this scale, ice melts at 32°, and water boils at 212°. Normal body temperature is 98.6° on the Fahrenheit scale or 37.0° on the centigrade scale. To convert from one scale to the other, the following equations are used:

$$(^\circ F) = \tfrac{9}{5}(^\circ C) + 32;$$

$$(^\circ C) = \tfrac{5}{9}[(^\circ F) - 32]$$

Appendix B

EXPONENTIAL NOTATION

THE NUMBERS encountered in the study of chemistry are often very large or very small. The usual decimal notations become cumbersome when one deals with such very large or small numbers, and it is convenient to use a more compact exponential notation.

The reader will recall that:

$$100 = 10 \times 10 \quad \text{or } 10^2 \text{ (two zeros)}$$
$$1000 = 10 \times 10 \times 10 \quad \text{or } 10^3 \text{ (three zeros)}$$
$$1{,}000{,}000 = 10 \times 10 \times 10 \times 10 \times 10 \times 10 \quad \text{or } 10^6 \text{ (six zeros)}$$

and so forth. In exponential notation, the unwieldy number is rewritten so that only one digit appears to the left of the decimal point, and the new number is then multiplied by 10 raised to the correct power. For example, if the population of a city is 8,200,000 people, we may first rewrite this number as $8.2 \times 1{,}000{,}000$, so that only one digit (the eight) appears to the left of the decimal point. Next, the number 1,000,000 is written in exponential form as 10^6, and we obtain the result:

$$8{,}200{,}000 = 8.2 \times 10^6$$

To use an example from the field of chemistry, there are known to be 602,000,000,000,000,000,000,000 molecules in one gram molecular weight. Obviously, this number is written more easily and apprehended more readily in the exponential form: 6.02×10^{23}.

The same rule applies for very small numbers, but now the exponent is negative. The reader will recall that:

$$0.1 = 10^{-1}$$
$$0.01 = 10^{-2}$$
$$0.001 = 10^{-3}$$

and so forth. Thus, if the diameter of an atom is known to be 0.000,000,030 cm, we first write this number as $3.0 \times 0.000,000,010$ cm, and finally in exponential notation as 3.0×10^{-8} cm.

Appendix C

GLOSSARY OF SCIENTIFIC TERMS

ABSORPTION in spectroscopy, the removal of light of a particular wavelength.

ACID a substance that can act as a hydrogen-ion donor.

ACTIVATED COMPLEX the unstable intermediate state with the highest energy in a chemical reaction.

ADSORPTION the binding of molecules onto the surfaces of substances.

ALPHA PARTICLE subatomic particle of mass 4 and charge +2; the helium nucleus.

ATOM fundamental building block of matter; atoms combine to form molecules.

ATOMIC MASS NUMBER atomic weight of an isotope, rounded off to the nearest integer.

ATOMIC NUMBER the net positive charge on the *nucleus* of an atom; equal to the number of *protons* in the nucleus.

ATOMIC WEIGHT the ratio of the weight of a sample of the natural element to that of a sample of carbon-12 containing an equal number of atoms, on a scale on which the atomic weight of carbon-12 is 12.00000.

AVOGADRO'S NUMBER the number of molecules in one gram molecular weight; 6.02×10^{23}.

BASE a substance that can act as a hydrogen-ion acceptor.

BETA PARTICLE an *electron* emitted by the nucleus of an atom in the process of radioactive decay.

BOND the link joining two atoms in a molecule.

BROWNIAN MOTION the irregular zigzag motion of tiny particles suspended in a liquid, owing to the impacts of molecules of the liquid on these particles.

CALORIE (gram calorie) a unit of energy; the amount of heat required to raise the temperature of one gram of water from 14.5 °C to 15.5 °C.

CATALYST a substance, other than a reactant, added to a chemical reaction in order to change its rate.

CATHODE RAY a beam of high-speed electrons emitted by the cathode (negative electrode) in a vacuum tube.

CIS *see* isomer, geometrical.

COMPOUND a pure substance formed by the chemical combination of two or more elements.

CONDUCTIVITY, ELECTRICAL a measure of the ability of a substance to allow the passage of an electric current.

COSMIC RAYS a highly penetrating type of radiation coming from outer space.

DECOMPOSITION takes place when a substance reacts, frequently at high temperature and/or reduced pressure, to form two or more new substances consisting of smaller molecules.

DENSITY mass per unit volume of a substance.

DIFFRACTION the modification that light or other radiation undergoes in passing through a narrow aperture or around the corner of a sharp object. A series of bright and dark bands are formed.

ELECTRIC CHARGE charge acquired by a body when it gains or loses electrons.

ELECTROLYSIS the decomposition of a compound through the agency of an electric current.

ELECTROMAGNETIC RADIATION a radiation emitted by oscillating electric charges; the *frequency* of the oscillation determines the nature of the radiation; in order of increasing frequency, the types of electromagnetic radiation are radio waves, infrared, visible, ultraviolet, and X rays or gamma rays.

ELECTRON a subatomic particle having a mass of 1/1840 of the mass of the hydrogen atom and a charge of -1.

ELECTRON VOLT a unit of energy; the energy of an electron when accelerated through a potential difference of one volt.

ELEMENT a pure substance that cannot be decomposed by chemical means.

ENERGY a concept that encompasses both heat and work; each substance under a given set of conditions has a characteristic amount of energy; a high-energy substance is likely to be highly reactive.

ENTROPY a measure of submicroscopic orderliness; a high entropy signifies a high degree of disorder.

EXCITATION the elevation of an atom or molecule from one energy state to another state of higher energy.

EXPONENT the number n in the expression x^n, which indicates that x is repeated as a factor n times.

FORMULA in chemistry, a graphical way of representing on paper the structure and composition of molecules. The *molecular formula* gives only the numbers of atoms of different kinds in one molecule; the *structural formula* shows also the manner in which the atoms are linked; the *geometrical formula* describes their arrangement in space.

FREQUENCY the number of times that a periodic event occurs in unit time. Also, the number of waves that pass a fixed point in a given interval of time.

GAMMA RAYS *electromagnetic radiation* of very high frequency emitted by the nuclei of atoms undergoing radioactive decay.

GRAM ATOMIC WEIGHT the weight of 6.02×10^{23} atoms of a given element; numerically equivalent to the atomic weight expressed in grams.

GRAM MOLECULAR VOLUME the volume occupied by one gram molecular weight of any gaseous compound at 0 ° C and 1 atm pressure; about 22.4 liters.

GRAM MOLECULAR WEIGHT the weight of 6.02×10^{23} molecules of a given compound; numerically equivalent to the molecular weight expressed in grams.

HETEROGENEOUS refers to a sample of matter of nonuniform composition.

HOMOGENEOUS refers to a sample of matter which is uniform throughout.

HYBRIDIZATION the mixing or combination of atomic orbitals to make a new set of orbitals with different properties.

INFRARED RADIATION radiation of wavelengths between 8×10^{-5} cm and 0.03 cm.

INSULATOR, ELECTRICAL a substance that does not readily conduct electric current.

ION an electrically charged atom or molecule.

ISOMERS two structurally different molecules having the same molecular formula. *Structural isomers* are molecules with the same molecular formula but different structural formulas. *Geometrical isomers* are molecules with the same structural formula but different geometrical formulas. The *cis*-isomer has two identical groups or atoms on the same side of the double bond; the *trans*-isomer has the identical groups on opposite sides of the double bond.

ISOTOPES two atoms of the same element (having the same atomic number) with different atomic mass numbers.

KINETIC ENERGY the energy possessed by a body by virtue of its motion.

LATTICE, CRYSTAL an orderly three-dimensional array of atoms, ions, or molecules. (See p. 218.)

LAW, SCIENTIFIC a generalization, based on experiments or observations, summarizing some phenomenon of nature or natural mode of behavior.

Laws are expressions of our factual knowledge and are not to be confused with *theories*.

MASS NUMBER *see* atomic mass number.

MOLE short for *gram molecular weight;* Avogadro's number of molecules.

MOLECULAR FORMULA *see* formula.

MOLECULAR WEIGHT the sum of the atomic weights of the constituent atoms in the molecule.

MOLECULES the smallest particles into which a pure substance may be subdivided, the collection of which still has the same properties as the original sample.

NEUTRON a subatomic particle of *mass number* one and zero charge.

ORBITAL see wave function. The term "orbital" is used also to denote pictures of the "95% probability boundaries" for the position of the electron.

OXIDATION the loss of electrons by an atom or molecule.

PHASE a homogeneous, physically distinct portion of matter, e.g., the liquid phase, the gas phase.

PHOTON an elementary particle of light; also a quantum of radiant energy.

POSITRON a subatomic particle having the same mass as the *electron,* but a charge of $+1$.

POTENTIAL ENERGY the energy possessed by a body due to its interaction with other bodies.

PROTON a subatomic particle of *mass number* one and a charge of $+1$.

QUANTUM NUMBER an integer that characterizes the quantized energy state of an atom or molecule.

REDUCTION the gain of electrons by an atom or molecule.

REFRACTIVE INDEX a measure of the speed at which light travels in the given sample of matter.

SPECTRUM a band of colors or a set of colored lines, formed when the light from a luminous object passes through a prism.

SYNTHESIS the art or process of making or building up a compound.

THEORY a convenient model, designed to explain a set of related observations or phenomena.

TRANS *see* isomer, geometrical.

ULTRAVIOLET RADIATION radiation of wavelengths between 1×10^{-6} cm and 4×10^{-5} cm.

WAVE FUNCTION a mathematical expression that is a solution of the wave equation. It defines the distribution of the electron in the vicinity of the atomic nucleus. Such expressions are also called atomic orbitals.

WAVELENGTH the distance between two adjacent identical points on a wave, e.g., the distance from peak to peak or trough to trough (λ).

Index

Index